Gary
God Bless You

Buddy

# OUR INCREDIBLE CREATED WORLD

**Ed Landry**
Missionary/Graphic Artist

**Buddy Petty**
Professional Engineer

*Don't tell me to have a nice day!*

**Our Incredible Created World**
It's no accident!

Authors - Ed Landry and Buddy Petty

Special thanks to our dedicated editorial team - Marcia Petty, Janet Landry. Tom and Janet Tucker, Marilyn Williams, and Rachel Butler.
.
Book and cover design by Ed and Janet Landry.
Illustrations by Ed Landry.

All images in the book are used by permission from:
Shutterstock, and Depositphotos.

ISBN-13: 979-8-9881173-0-8

Printed in China
CHINA SEVEN COLOR GROUP CO. LTD
www.chinaprinting4u.com

`

**CONTACT INFORMATION**
Website: upliftingchristianbooks.com
Contact: Ed@upliftingchristianbooks.com

Uplifting Christian Books
Nashville, Tennessee
2023

# Dedication

Today we have access to very high-powered microscopes and information that previous generations did not have. We understand more today about how our highly complex world is interconnected.

How did the universe begin? Some believe it was formed by undirected natural forces. Evolution is one explanation of how it all started beginning with nothing, and then over billions of years and billions of small beneficial changes, life developed on its own into what we see today. The agent of all the change is what they call natural selection.

Others see the work of an eternal God who created it all from nothing. "In the beginning God created the heavens and the earth." Mankind was the crowning act of God's handiwork. We alone are made in His moral image and capable of having a relationship with the Creator of all things.

In this book, you will see 50 different examples of the natural world and their complex functions. The facts presented are agreed upon by both evolutionists and creationists. This book lays out the information. What you do with it is up to you.

This book is dedicated to all our children and grandchildren. May you find a new sense of wonder and joy as you discover the designs and art of the world's greatest living Artist, God!

*"I will give thanks to You, for I am fearfully and wonderfully made; Wonderful are Your works,"* Psalm 139:14

# CONTENTS

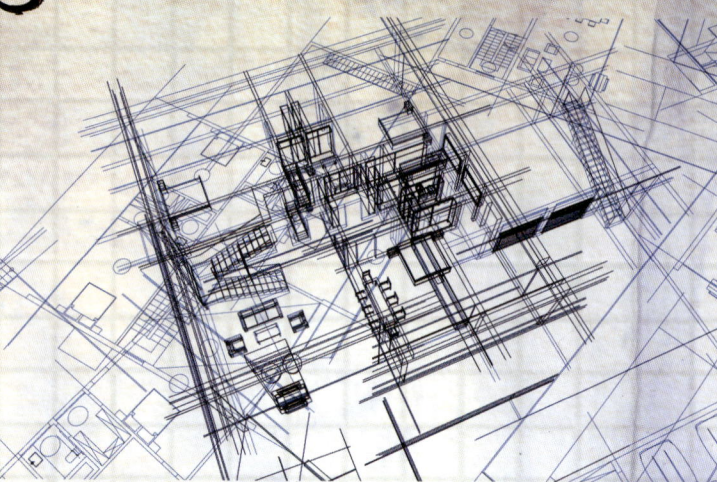

Imagine a high rise office building. It took architects, engineers, and builders to make it happen. Design means there is a designer.

Did you know that a single cell in your body is over a million times more complex than a high-rise building? That includes all the computers and electronics combined.

Now ask yourself. Did that living cell just happen by accident? Or, did an all powerful God make that cell?

# Understanding EVOLUTION

## Here is a simple overview of the theory of Natural Evolution.

• Before anything existed there was nothing, absolutely nothing.

• Then something came into existence from nothing. No one knows exactly how!

• That first something is called the "singularity."

• The singularity is described as a very, very, tiny dot of highly condensed stuff that exploded and became the entire universe. This happened 14 billion years ago. No one knows how that all happened.

• Over billions of years, the hot rocks from the "big bang" explosion cooled off forming our universe.

• For the next billions of years the various elements on earth combined and formed swampy pools from which the first life began.

• Through various mutations and natural selection the life forms changed slowly and in small increments. All life we have on earth today is a result of accidents, mutations, natural selection, and time.

• All life has come from other life.

# Evolutionists believe all life originally evolved from a dead rock!

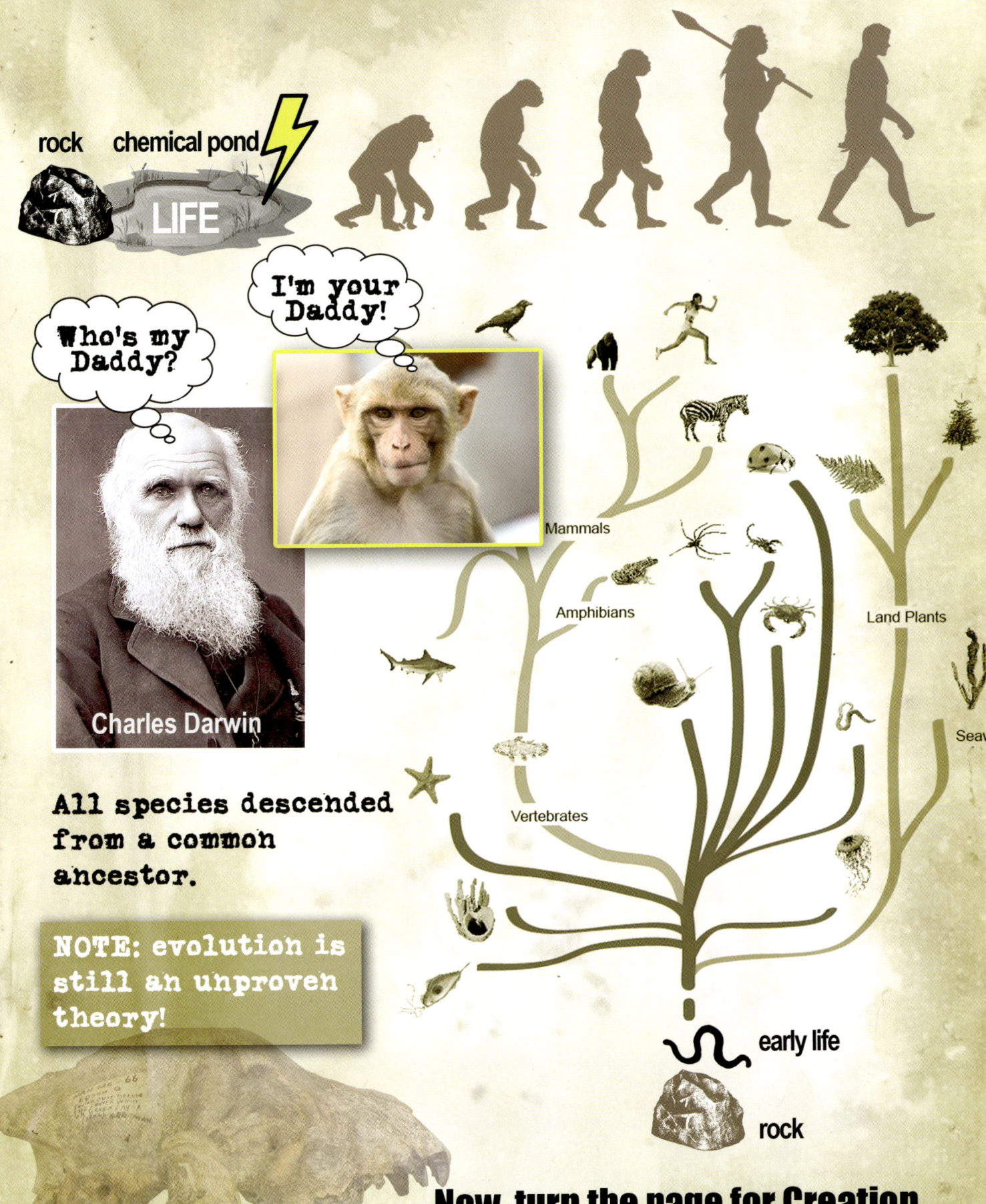

rock   chemical pond   LIFE

Who's my Daddy?

I'm your Daddy!

Charles Darwin

All species descended from a common ancestor.

NOTE: evolution is still an unproven theory!

Mammals

Amphibians

Land Plants

Vertebrates

Seawe

early life

rock

**Now, turn the page for Creation.**

# CREATION

Does the design and complexity of the natural world provide proof that this was not an accident, but the work of an infinite God?

*"In the beginning God created the heavens and the earth."*

Genesis 1:1

This book has 50 examples of our complex and beautifully designed world. Design requires a Great Designer.

Evolution provides no purpose or hope. With God, there is eternal hope and wonderful purpose in life.

Now it is up to you to make your own decisions about how you believe this all happened.

*"For since the creation of the world His invisible attributes, His eternal power and divine nature, have been clearly seen, being understood through what has been made, so that they are without excuse."*

Romans 1:20

# A BLADE OF GRASS
## under the microscope

A high power microscope took this picture of the cells in a blade of grass.

Cells are like building blocks with different jobs. Every blade of grass is made of millions of cells.

Let's take a deeper look at one of the grass cells.

10

**Each of the millions of grass cells is like a highly complex city.**

Before you look at the next page, remember every component within the complex cell is necessary for the cell to survive. If cells don't grow, the grass does not grow. Green plants produce oxygen so we can breathe. Everything is interconnected. It all has to work perfectly. On the next page you will see how amazingly complex this simple blade of grass is.

# A BLADE OF GRASS

## ...continued

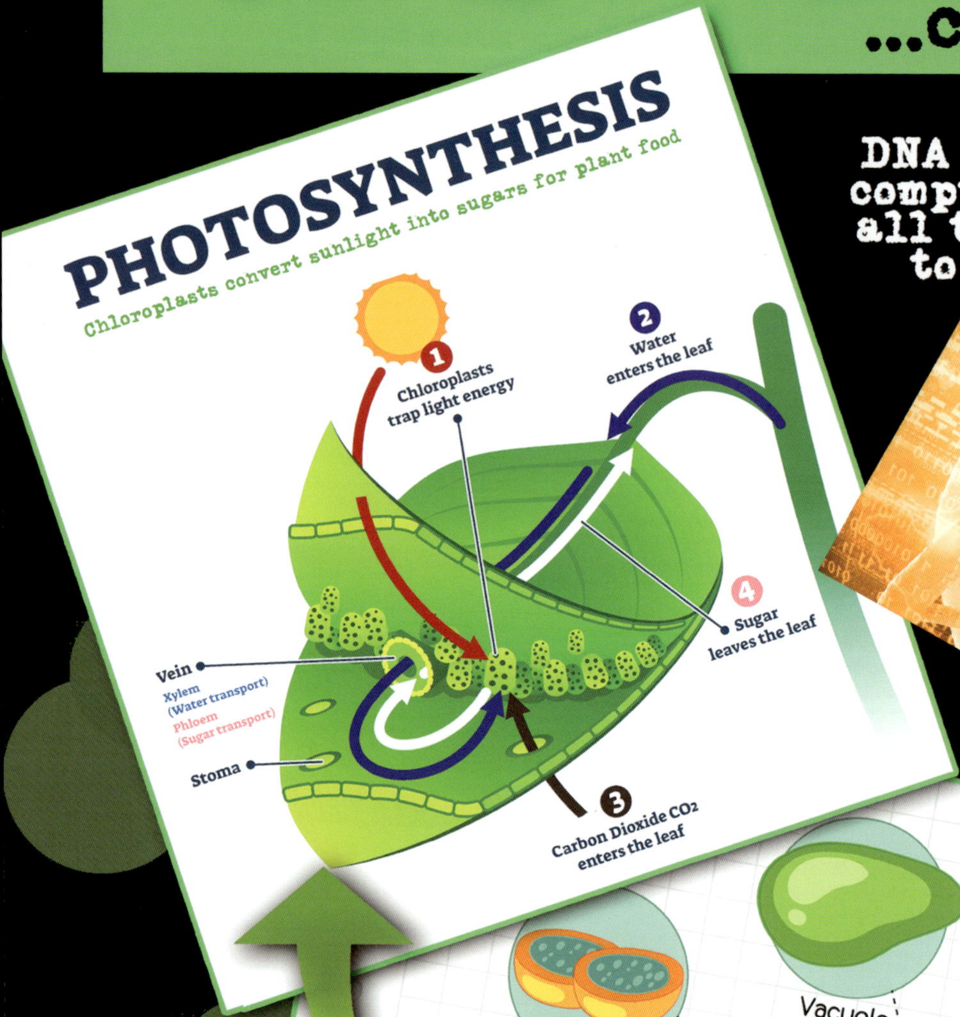

PHOTOSYNTHESIS
Chloroplasts convert sunlight into sugars for plant food
1 Chloroplasts trap light energy
2 Water enters the leaf
3 Carbon Dioxide $CO_2$ enters the leaf
4 Sugar leaves the leaf
Vein
Xylem (Water transport)
Phloem (Sugar transport)
Stoma


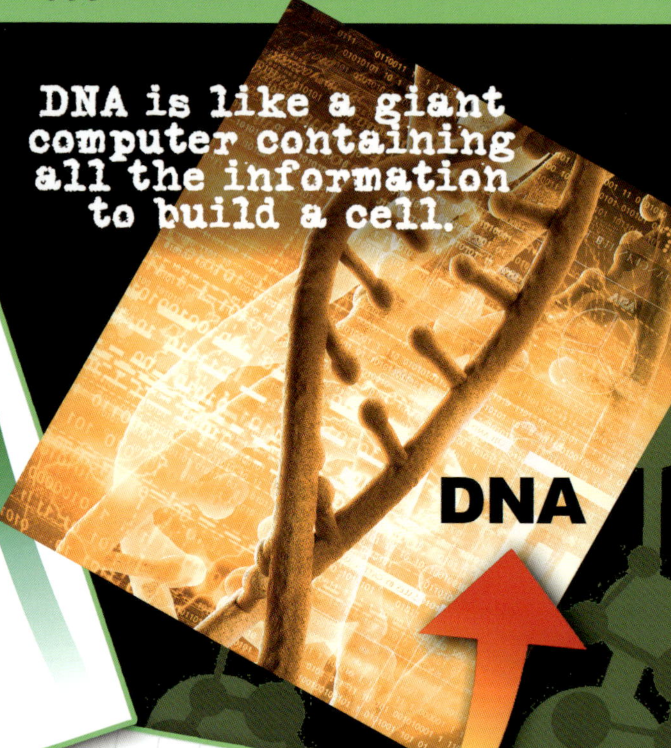

DNA is like a giant computer containing all the information to build a cell.
DNA


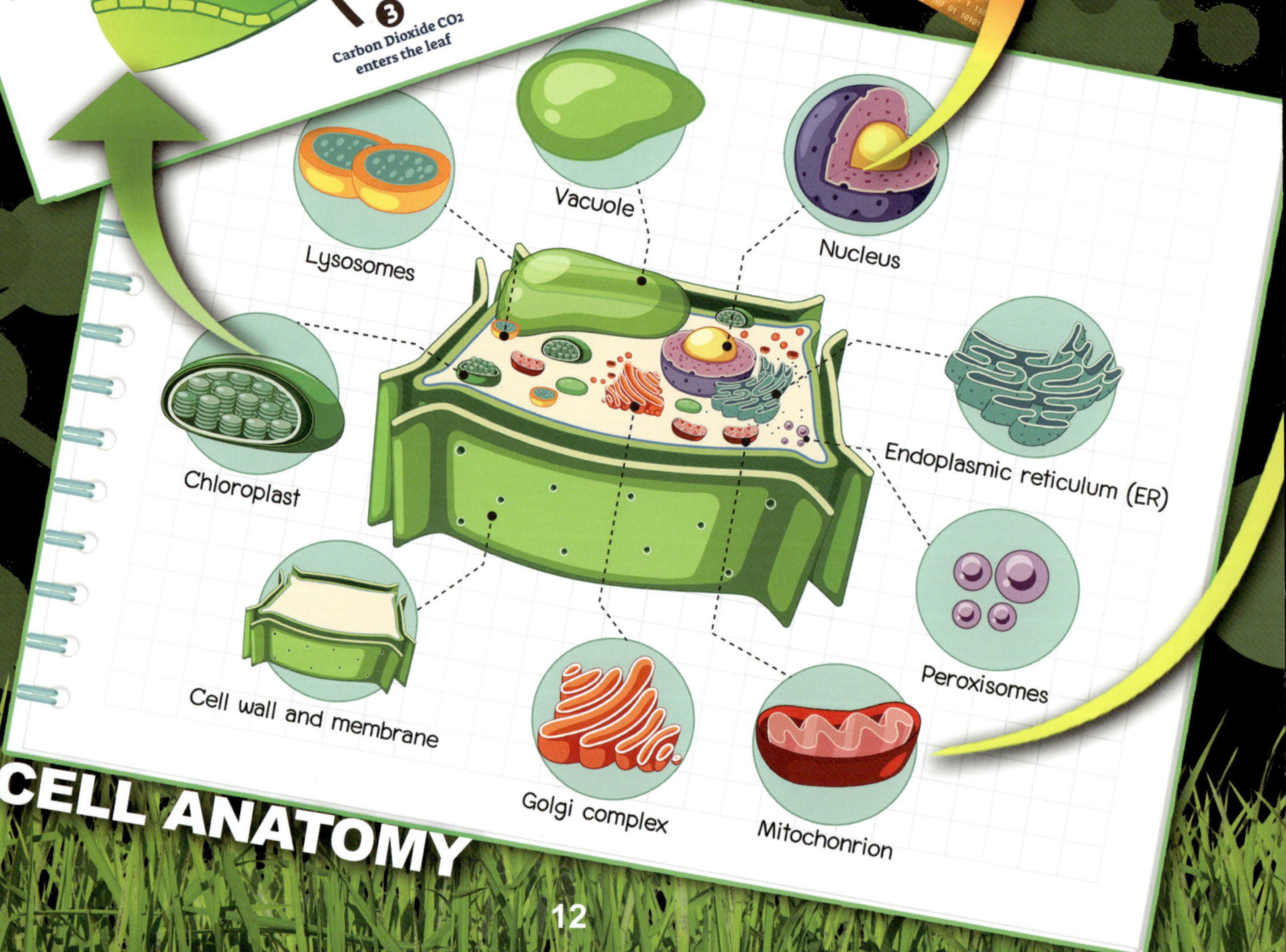

Vacuole
Lysosomes
Nucleus
Chloroplast
Endoplasmic reticulum (ER)
Peroxisomes
Cell wall and membrane
Golgi complex
Mitochonrion


CELL ANATOMY

Each of the millions of cells contains hundreds of mitochondria which are the power plants of the cell. They generate electrical energy. These miniature power plants have rotating, multistage generators. Each one is a masterpiece of design.

If we go deeper yet, we come to the atoms that are like miniature universes with orbiting electrons, but that is another story. The complexity and design of a blade of grass is beyond imagination.

There are 100,000,000,000,000 (100 trillion) atoms in a single cell of a simple blade of grass! Don't forget, this is just one cell!

# BAR-TAILED GODWIT
## world's most amazing migration

The four-month-old bar-tailed godwit is the only bird in the world that can fly 1/3 of the distance around the earth without stopping!

### HOW DOES IT DO THAT?

**To** undertake the longest non-stop flight, it needs a lot of fuel. In Alaska the godwit feeds on fatty foods and before they leave on their long journey, they carry the greatest fat loads of any migratory bird. They are able to accomplish this by reducing the size of their digestive organs to make room for the fat storage. It is like adding extra fuel tanks to a plane.

They also have the ability to predict favorable weather and wind patterns.

Did you know there is another bird that does this, just not quite as far. It is the Pacific Golden Plover.

Look it up.

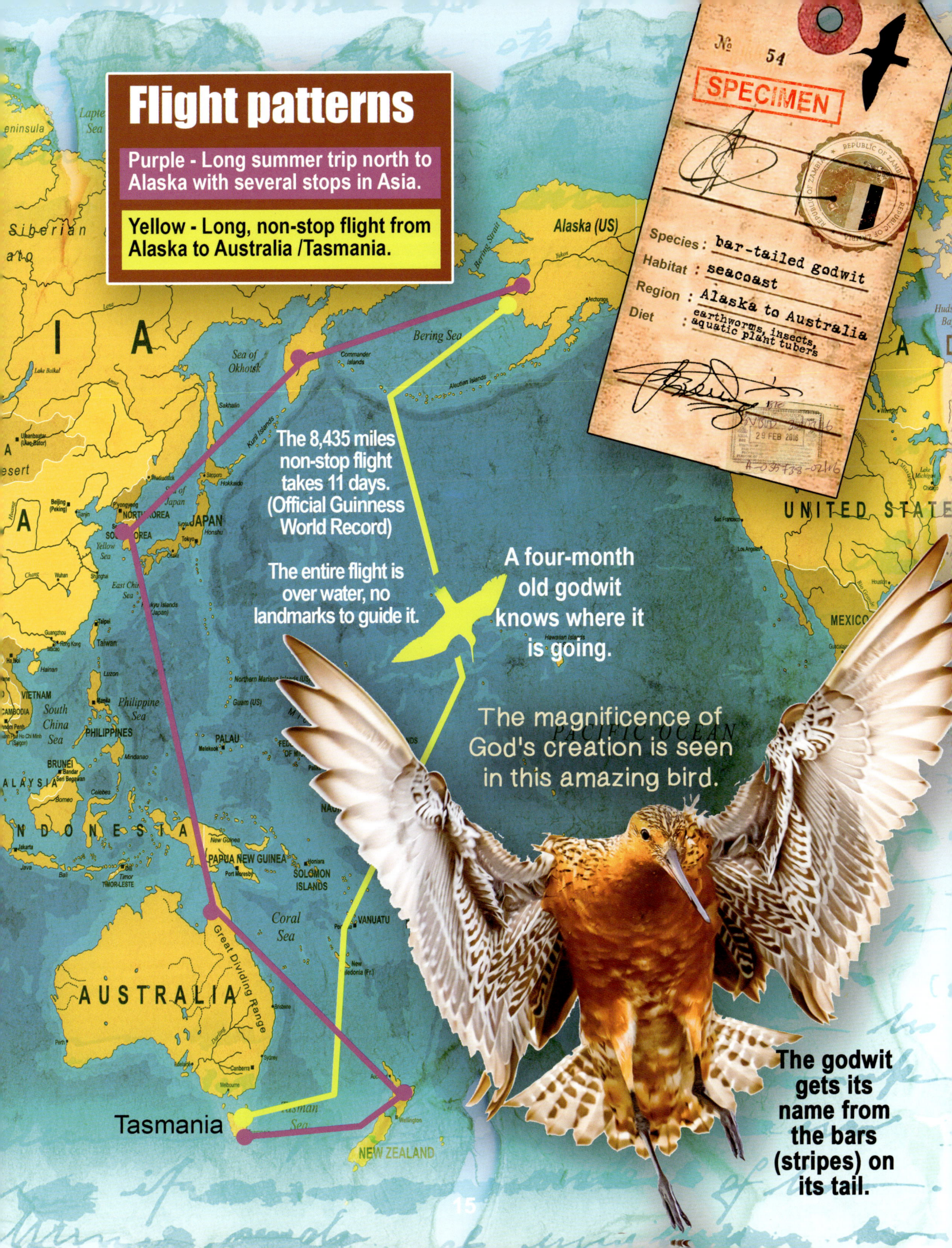

# Flight patterns

**Purple** - Long summer trip north to Alaska with several stops in Asia.

**Yellow** - Long, non-stop flight from Alaska to Australia /Tasmania.

No. 54

SPECIMEN

REPUBLIC OF ZAMBIA

Species : bar-tailed godwit
Habitat : seacoast
Region : Alaska to Australia
Diet : earthworms, insects, aquatic plant tubers

29 FEB 2016

The 8,435 miles non-stop flight takes 11 days. (Official Guinness World Record)

The entire flight is over water, no landmarks to guide it.

A four-month old godwit knows where it is going.

The magnificence of God's creation is seen in this amazing bird.

The godwit gets its name from the bars (stripes) on its tail.

Tasmania

# THE ANGLERFISH
## the fish with a fishing pole

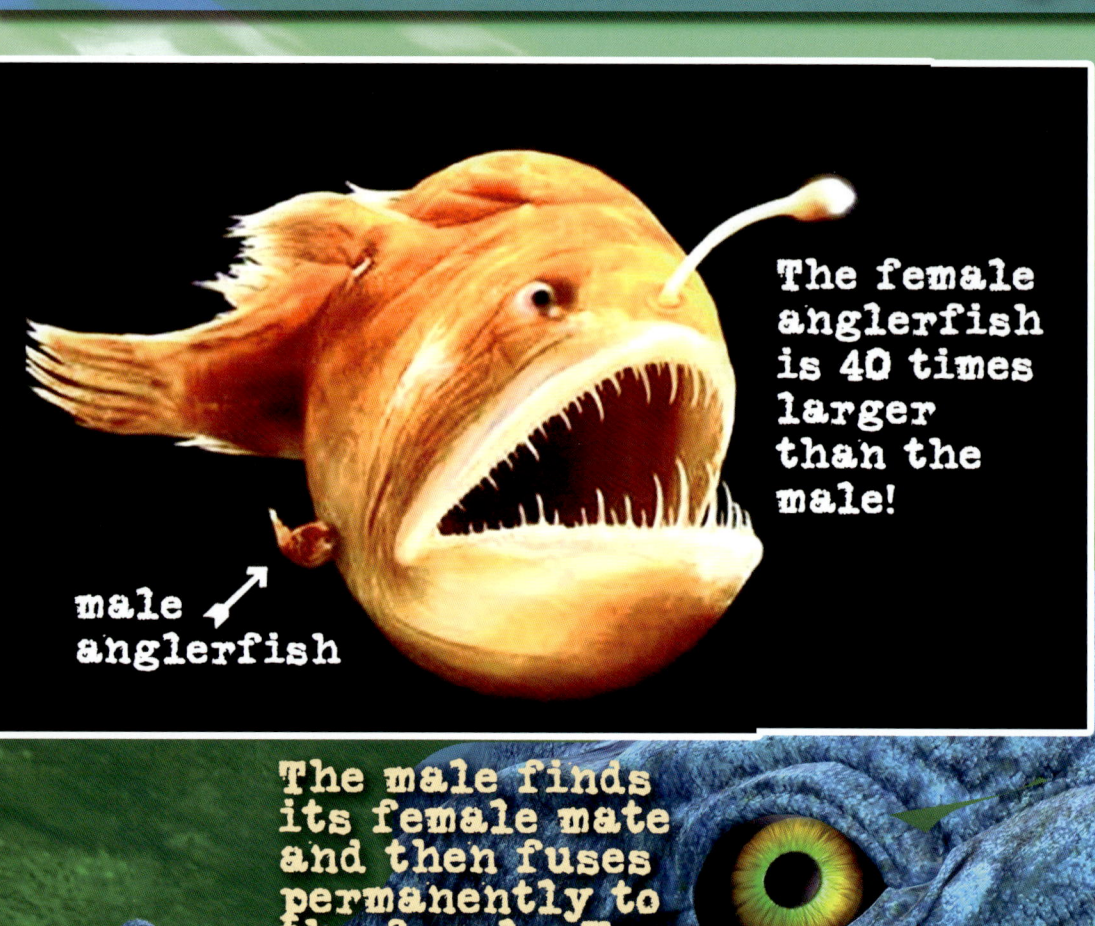

The female anglerfish is 40 times larger than the male!

male anglerfish

The male finds its female mate and then fuses permanently to the female. From then on they both share the same circulatory system.

**Anglerfish**
teleost order lophiiformes

Diet: fish
Life Span 30 years
Size: up to 3 feet
Habitat: deep sea

SPECIMEN

It opens its mouth so wide it can swallow a fish twice its size.

At the end of the fishing rod that grows out of the head of the anglerfish is a lure that attracts its next meal. It glows in the dark.

The source of this light is bacteria that have an ability to produce light. Fish are attracted to the fleshy tip used as bait. The bacteria and the anglerfish have a mutual (symbiotic) relationship. The bacteria are unable to create the light unless connected to the anglerfish. They need each other to survive.

The Anglerfish species have more than 200 subspecies, and the majority of them live in the depths of the Antarctic and Atlantic oceans.

This fish has so many unique features that the only explanation of its origin is Divine design and creation. To imagine that the anglerfish evolved from another fish raises too many questions that have no answers.

**?**

## WHO DO YOU THINK WOULD WIN AN UGLY FISH CONTEST?

**Anglerfish**

**Blobfish**

# THE HUMMINGBIRD

## the bird that acts like a helicopter

The first helicopter was designed by Leonardo DaVinci around A.D. 1480. It never worked.

## FACTS

Hummingbirds are the tiniest birds in the world. They are also the smallest migrating bird. Unlike the large flocks of other species, hummingbirds typically travel alone for up to 500 miles at a time. Some travel 4,000 miles from Mexico to Alaska every spring.

Hummingbirds are the only birds that can fly like a helicopter: up, down, sideways, forward, backwards, and even upside-down. No other bird can match the hummingbird for agility. Their wings move in a figure eight pattern, which allows them to maneuver with ease. Typical bird flight is achieved by flapping the wings up and down. Hummingbirds, however, rotate or twist their upper arm bones to invert their wings and gain lift from the upstroke as well as the downstroke. The result? It's a fact that hummingbirds are the most agile birds on the planet. They beat their wings more than 50 times per second, and even faster in extreme flight mode.

Hummingbirds have a great memory. They remember every flower and feeder they've been to, and how long it will take a flower to refill.

Unlike other birds, the shoulders are ball and socket joints that allow the hummingbird to rotate their wings one hundred eighty (180) degrees in all directions.

THERE ARE 325 UNIQUE KINDS OF HUMMINGBIRDS.

18

## ¿DID YOU KNOW?

A hummingbird is the only bird in the world that can fly backward.

The average weight of a hummingbird is less than a nickel. Their tiny legs are only used for perching and moving sideways while perched. They can't walk or hop.

Hummingbirds drink the nectar found in feeders by moving their tongues in and out about 13 times per second.

A hummingbird breathes 250 times a minute and its heart beats up to 1,260 times per minute.

No other bird is like the humming bird. It did not evolve from another bird.

Hummingbirds fly at an average of 25-30 miles per hour and can dive up to 50 miles per hour.

They beat their wings between 50 and 200 flaps per second.

Hummingbird bones are extremely porous. Bones in the wings and legs are hollow to save even more weight.

Red is the favorite color to attract hummingbirds.

# THE ELECTRIC FISH COMPANY
## creatures that will shock you!

650 VOLTS

Electric eels aren't actually eels. They're members of the knife fish family.

A fully grown electric eel can be up to eight feet long and weigh 44 pounds!

They don't spend all their time underwater. They have to come to the surface to breathe.

The eel's vital organs are in the front 20 percent of its body. The rest is packed with 6,000 cells that act like tiny batteries that produce a 650 volt charge..

The eels are nearly blind and use a radar-like system of electrical pulses to navigate and find food.

Only God could create a fish that generates high-voltage electricity.

Go ahead and make my day!

The 400 volt catfish.

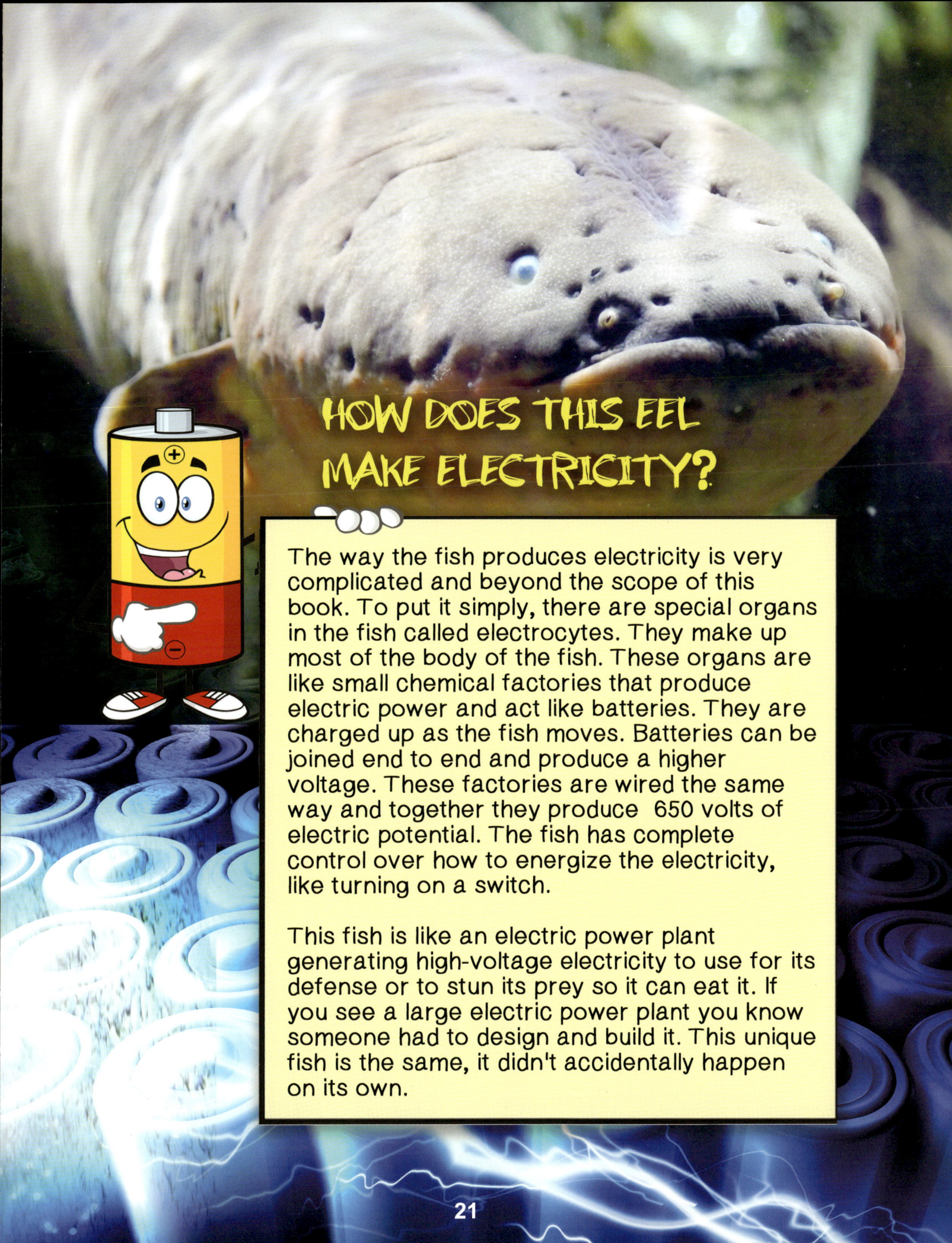

# HOW DOES THIS EEL MAKE ELECTRICITY?

The way the fish produces electricity is very complicated and beyond the scope of this book. To put it simply, there are special organs in the fish called electrocytes. They make up most of the body of the fish. These organs are like small chemical factories that produce electric power and act like batteries. They are charged up as the fish moves. Batteries can be joined end to end and produce a higher voltage. These factories are wired the same way and together they produce 650 volts of electric potential. The fish has complete control over how to energize the electricity, like turning on a switch.

This fish is like an electric power plant generating high-voltage electricity to use for its defense or to stun its prey so it can eat it. If you see a large electric power plant you know someone had to design and build it. This unique fish is the same, it didn't accidentally happen on its own.

# GRASSHOPPERS AND LOCUSTS
## nature's Incredible Hulk

The Incredible Hulk is a popular movie about a man named Bruce Bannon who transforms into a physically different person when he gets upset and his heart rate races to over 200 beats per minute. Of course, this is a fictional story, but grasshoppers undergo a similar transformation in real life.

## What causes them to change into monsters?

Grasshoppers are generally solitary insects who go about their business without bothering anyone.

A grasshopper becomes a locust only under certain environmental conditions. Heavy rains before a drought, can cause ordinarily solitary grasshoppers to become highly social. Contact with other grasshoppers stimulates the release of serotonin. Physical changes occur. The locusts become aggressive as their wings grow longer. They grow stronger and their color changes.

# LOCUSTS IN HISTORY

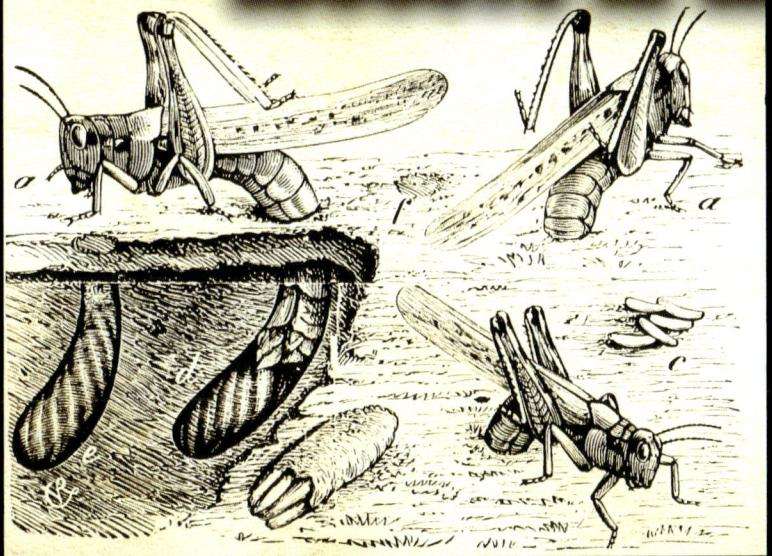

In North America in 1895, a swarm of Rocky Mountain locusts, measuring 1,800 miles long and 110 miles wide, blocked the sun for five days. It had an estimated 3.5 trillion locusts. Thirty years later, the locust was extinct!

In the Bible Moses warned Egypt's Pharaoh that God would send so many locusts that they would "cover each and every tree of the land and eat all that is there to be eaten." God used plagues like the locusts to free His people from slavery to go to the land of Israel which became their new homeland. (Exodus 10:1-20)

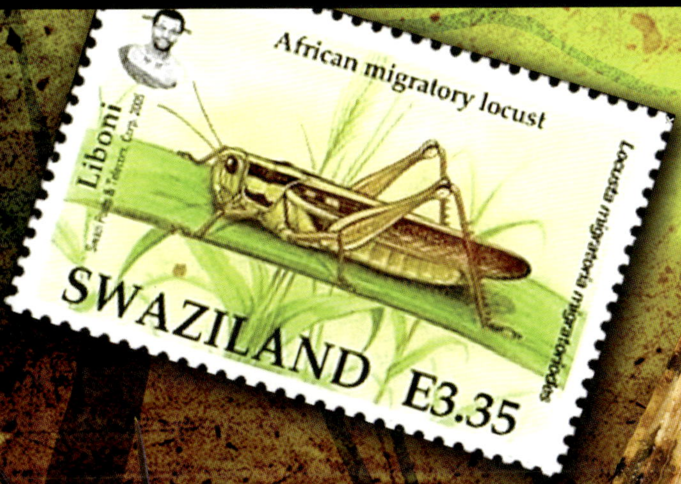

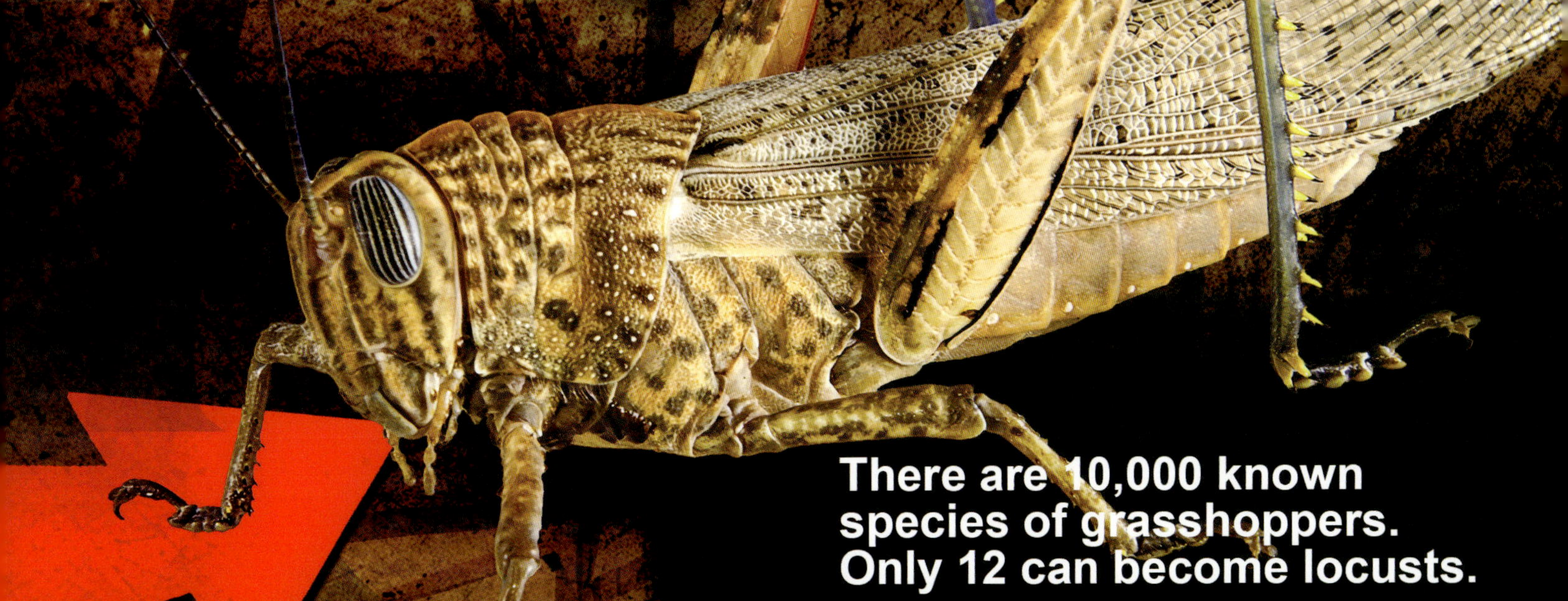

African migratory locust
Liboni
Locusta migratoria migratorioides
SWAZILAND
E3.35

There are 10,000 known species of grasshoppers. Only 12 can become locusts.

# DUCK-BILLED PLATYPUS
## the oddest creature on earth

The Duck-billed platypus is one of the strangest creatures in the world. It is as if God used spare parts from other animals He created to make the platypus. The platypus has a paddle-shaped tail, like a beaver; a sleek, furry body, like an otter; and a flat bill and webbed feet, like a duck. Their legs sprawl out to the side of the body, giving it a lizard-like walk. They lay eggs like a bird and are venomous like a snake.

**Ear slits**

**Closeable nostrils**

**Duck's bill**

The platypus does not use its eyes, ears, and nose when swimming underwater to search for food. They use unique electro-receptors located in the skin of their bill to detect small electrical signals from their prey

**Otter's fur**

They scoop up insects, larvae, shellfish, and worms in their bill, along with bits of gravel and mud. This material is stored in cheek pouches until the platypus is back on land. They do not have teeth, so the gravel bits help them "chew" their meal.

Only found on the east coast of Australia

AUSTRALIA
PLATYPUS
9d
POSTAGE

Flattened furry tail like a beaver

When it comes to eating, platypuses don't have a stomach – their esophagus connects directly to their intestines.

Poisonous spike on the male's ankle

Retractable claws

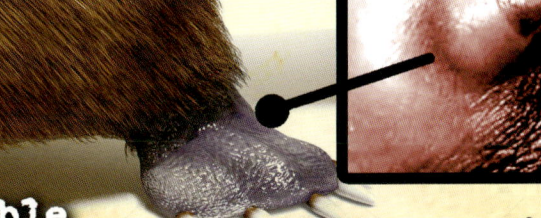

Venom like a snake

Beaver-like webbed feet

I am one of two mammals that lay eggs.

# THE ANT FARM

**1**

### FUNGUS FARMING

Leaves are gathered by ants and chewed into small pieces, then stacked together for fungus to digest. Ants consume the fungus.

**Ants feasting on fungus**

### MUTUAL BENEFITS

Everyone wins. The aphids provide a necessary food for the ants and the ants make sure the aphids have food and safety. God created an amazing, interconnected world. There are thousands of mutually dependent relationships like the ant and the aphid.

**5**

### MILKING OPERATION

Ants "milk" the aphids—stroking the aphid's stomach with their antennae, coaxing them to secrete their honeydew, which is then eaten up by the ant. It is like a farmer milking a cow for food.

**Milking an aphid**

## 2

### HONEYDEW FARMING

Aphids eat plants and then poop out a sugary liquid called honeydew, which is a favorite food for ants. Ants utilize a unique system of aphid farms to produce their needed honeydew.

Ant eating a honeydew droplet

Aphids

## 3

### THE FARMWORKERS

Ants gather and control colonies of aphids to harvest the honeydew. Aphids are their farm workers and the ants make sure the aphids have a continual supply of food to produce honeydew. They even move the colonies to new plants when they need more food source. They are herding aphids.

### FARM SUPERVISORS

The ants protect their worker aphids by killing off the enemies of aphids, like ladybugs. Sometimes the aphids will grow wings to fly away but the ants bite their wings off and even produce chemicals in their glands that cause the aphids to slow down. This keeps the aphids on the farm where they are safe and productive.

## 4

# HUMAN DNA
## the complicated code of life

"When it comes to storing massive amounts of information, nothing comes close to the efficiency of DNA. A single strand of DNA is thousands of times thinner than a strand of human hair. One pinhead of DNA could hold enough information to fill a stack of books stretching from the earth to the moon 500 times!"
- Answers in Genesis

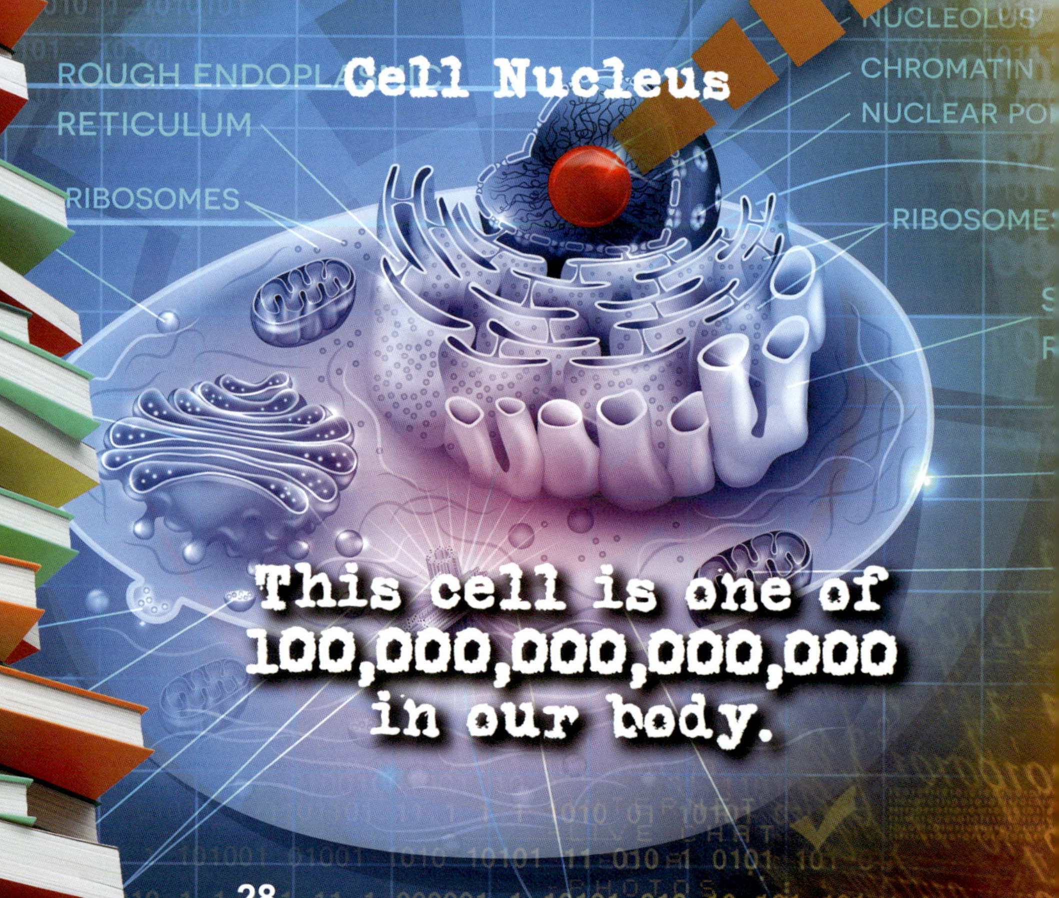

**Cell Nucleus**

ROUGH ENDOPL
RETICULUM

RIBOSOMES

NUCLEOLUS

CHROMATIN

NUCLEAR PO

RIBOSOME

This cell is one of 100,000,000,000,000 in our body.

# Cell Nucleus

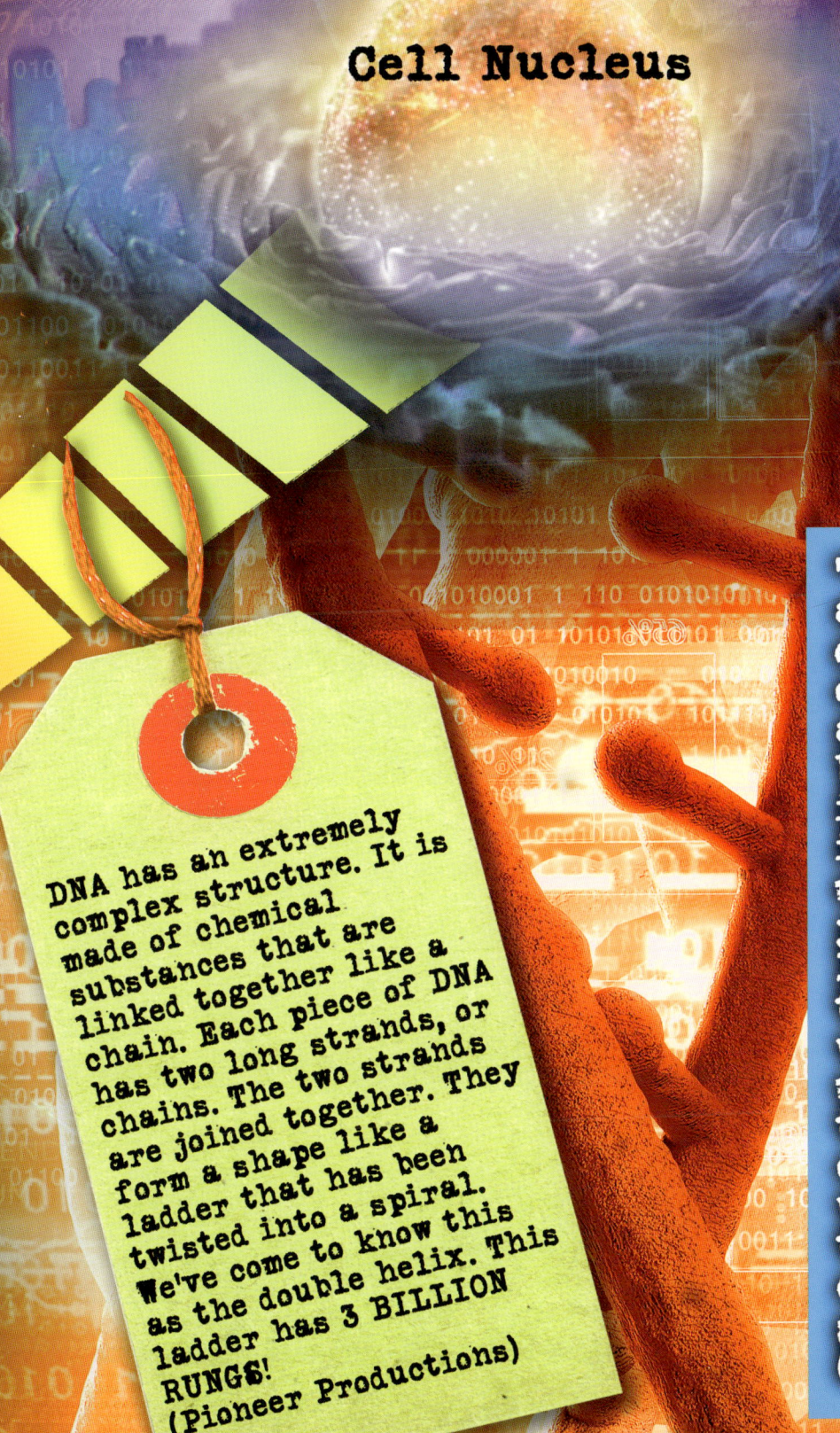

The cell nucleus is a safe vault that contains the DNA, the complete blueprints of our lives.

DNA has an extremely complex structure. It is made of chemical substances that are linked together like a chain. Each piece of DNA has two long strands, or chains. The two strands are joined together. They form a shape like a ladder that has been twisted into a spiral. We've come to know this as the double helix. This ladder has 3 BILLION RUNGS! (Pioneer Productions)

"Your DNA is arranged as a coil of coils of coils of coils of coils! This allows the 3 billion base pairs in each cell to fit into a space just 6 microns across. If you stretched the DNA of one cell all the way out, it would be about 2 meters long and all the DNA in all your cells put together would be about twice the diameter of the Solar System." (Science Focus)

There is simply no possible way for the structure and the information in the DNA to have just happened by accident. It had to be designed and only God could do that.

# THE OXPECKER
## Africa's odd couple

This little bird helps many of Africa's large, hoofed animals, including giraffe, antelope, zebra, Cape buffalo, and the rhinoceros.

## THEY BENEFIT EACH OTHER

The birds pick at parasites on the animal's body, including ticks and blood-sucking flies. By eating the ticks from the hide of the water buffalo, the oxpecker gets a free meal, and the water buffalo gets cleaned of parasites.

In addition, the oxpecker will eat diseased wound tissue, keeping wounds clean as they heal.

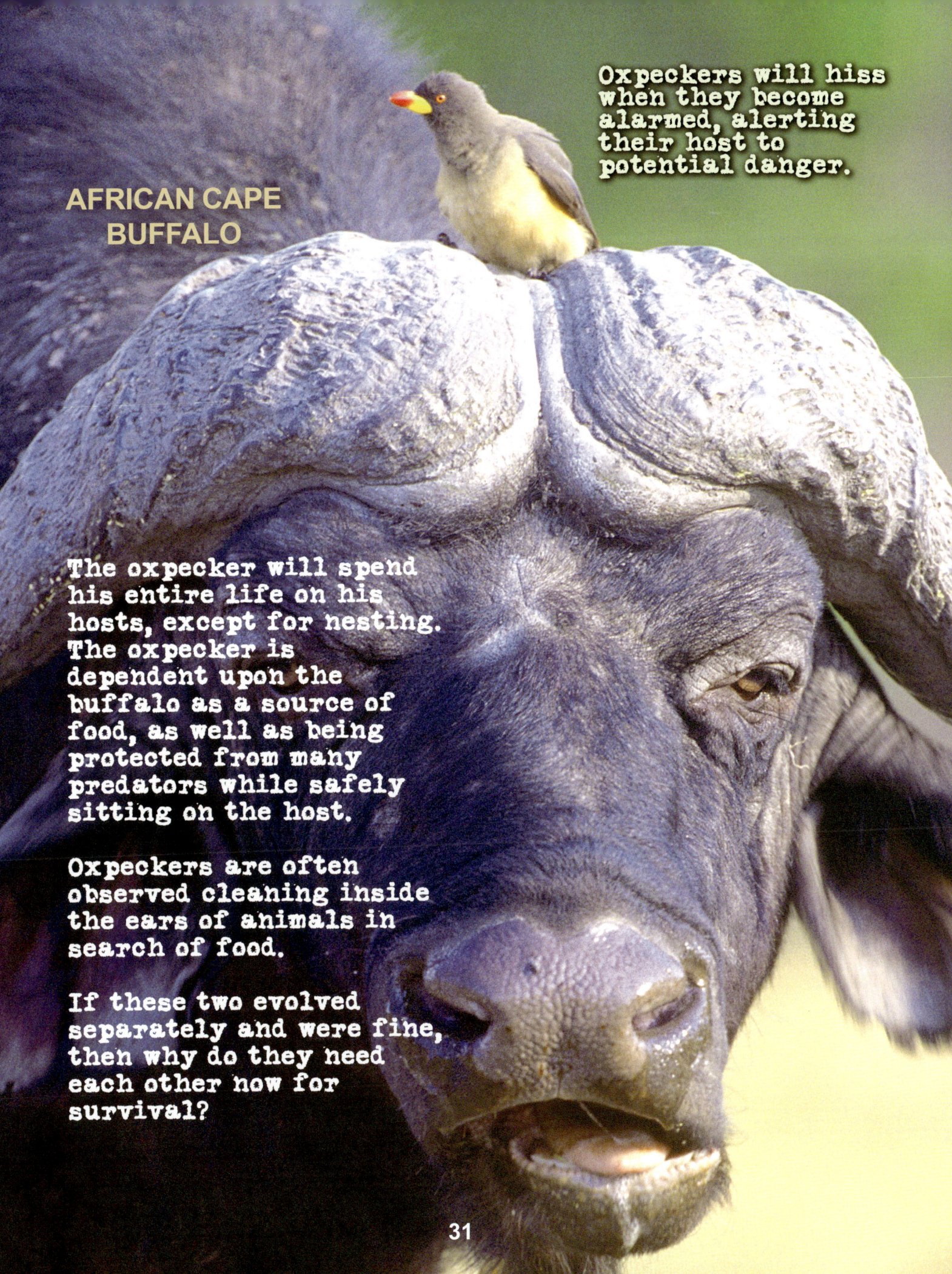

## AFRICAN CAPE BUFFALO

The oxpecker will spend his entire life on his hosts, except for nesting. The oxpecker is dependent upon the buffalo as a source of food, as well as being protected from many predators while safely sitting on the host.

Oxpeckers are often observed cleaning inside the ears of animals in search of food.

If these two evolved separately and were fine, then why do they need each other now for survival?

31

# THE TARDIGRADE
## the most indestructible creature on earth

**UNIQUE SPECIES**

They are smaller than the width of a a human hair.

What is this? Is it a stuffed bear? Is it an alien from space? No, it is a small animal you can only see under a microscope. People call them "water bears."

**1,300 SPECIES**

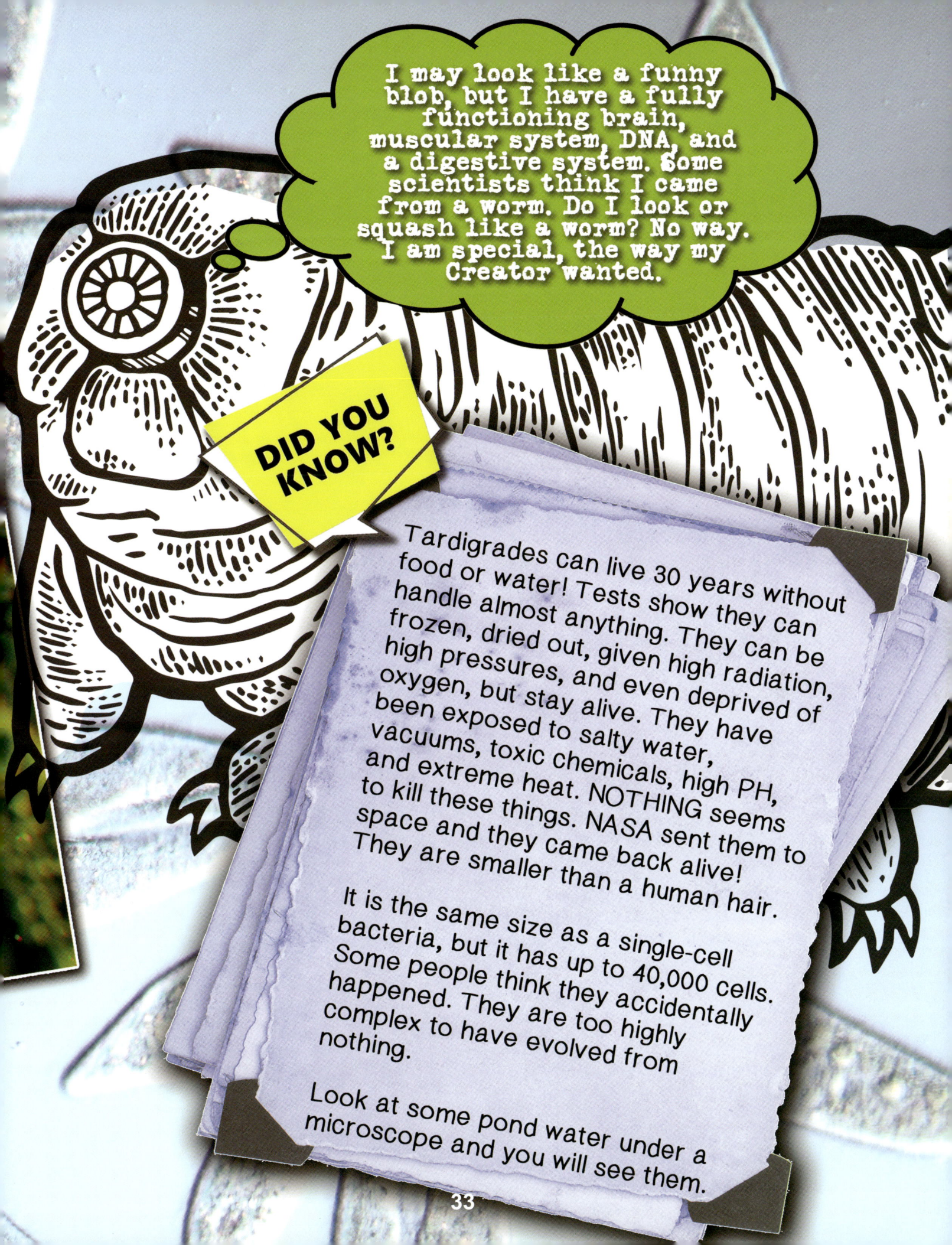

I may look like a funny blob, but I have a fully functioning brain, muscular system, DNA, and a digestive system. Some scientists think I came from a worm. Do I look or squash like a worm? No way. I am special, the way my Creator wanted.

**DID YOU KNOW?**

Tardigrades can live 30 years without food or water! Tests show they can handle almost anything. They can be frozen, dried out, given high radiation, high pressures, and even deprived of oxygen, but stay alive. They have been exposed to salty water, vacuums, toxic chemicals, high PH, and extreme heat. NOTHING seems to kill these things. NASA sent them to space and they came back alive! They are smaller than a human hair.

It is the same size as a single-cell bacteria, but it has up to 40,000 cells. Some people think they accidentally happened. They are too highly complex to have evolved from nothing.

Look at some pond water under a microscope and you will see them.

# VENUS FLYTRAP

## what's on the menu?

GOURMET CUISINE

**VENUS FLYTRAP CAFE**

menu

### MAIN COURSE

FAT FLIES

BEETLES

SPIDERS

### Catch of the day
### SOUP

SOWBUG BROTH WITH PLUMP BLOODWORMS

### DESSERT
Today's Smoothies

FRESH SLUGS

EARTHWORMS

### Fresh APPETIZERS

SMALL ANTS

BABY GRASSHOPPERS

CRUNCHY CRICKETS

Some insects require several weeks to digest.

Venus flytraps are the only plant that eats meat. They are native to South Carolina and North Carolina.

34

Teeth

Trigger Hairs

Nectar Secreting Glands

Digestive Glands

The Venus flytrap is one of a kind. Nectar in its "traps" attracts insects. When an unsuspecting insect brushes against tiny trigger hairs, the trap snaps shut in less than a second. Once the trap is tightly closed, digestive acids and enzymes dissolve the insect, and the plant absorbs the nutrient-rich "soup." Seven to ten days later, the trap opens, ready for another meal.

Charles Darwin called the Venus flytrap plant, "one of the most wonderful in the world." He wondered how the trapping mechanism could have evolved. The design and geometry is elegant. There is nothing else like it in the world.

It is one more evidence of Creation, not evolution.

## INTERESTING FACTS

Most plants take in nutrients from the soil. Venus flytraps aren't like most plants! They obtain most of the nutrients they need from the insects and other small wildlife they capture.

A trap's trigger hairs must be struck twice within a 20-second time frame for the trap to snap shut.

The trapping mechanism of a Venus flytrap is one of the fastest movements observed in the plant kingdom.

# FUNGUS AMONG US
## the internet of the soil
### molds, yeasts, and mushrooms

Fungi decompose dead organic matter, like trees, leaves, and plants. They are the garbage disposal agents of the world. Their tiny tentacles (mycellium) spread out under the forest floor taking vital nutrients to all the plants. They turn death into life. Without fungi, life on earth would not continue. They are like a giant underground life sustaining web.

They have been called the "wood-wide web" by scientists because they connect the entire ecosystem.

Honey
Mushroom

Researchers have measured up to 8 miles of fungi filaments in one teaspoon of soil.

# FUNGI ARE HIGHLY COMPLEX CELL STRUCTURES

Septum

Vacuole

Golgi apparatus

Plasma membrane

Cell wall

Mitochondrion

Lipid body

Nucle

## SINGLE FUNGI CELL

**DID YOU KNOW?**

Scientists estimate that there are over 5,000,000 species of fungi on earth, but we have only discovered about one percent of them.

They have a complex DNA.

The largest living organism in the world is a mushroom. The honey mushroom covers 2.4 miles in the Blue Mountains of Oregon.

## Important note:

The only way life can exist on earth is if the fungi are alive and present. For fungi to grow, every single part of their complex cell system must be complete. No part can be missing. Fungi are so complicated and perfectly designed, they had to be created. They could not have happened by accident.

# OUR AMAZING EYES
## can you see me now?

"To suppose that the eye, with all its incomparable apparatus for adjusting the focus to different distances, for admitting different amounts of light, and for the correction of spherical and chromatic deviation, could have been formed by natural selection, seems, I freely confess, absurd in the highest possible degree." – *Charles Darwin, Origin of the Species*

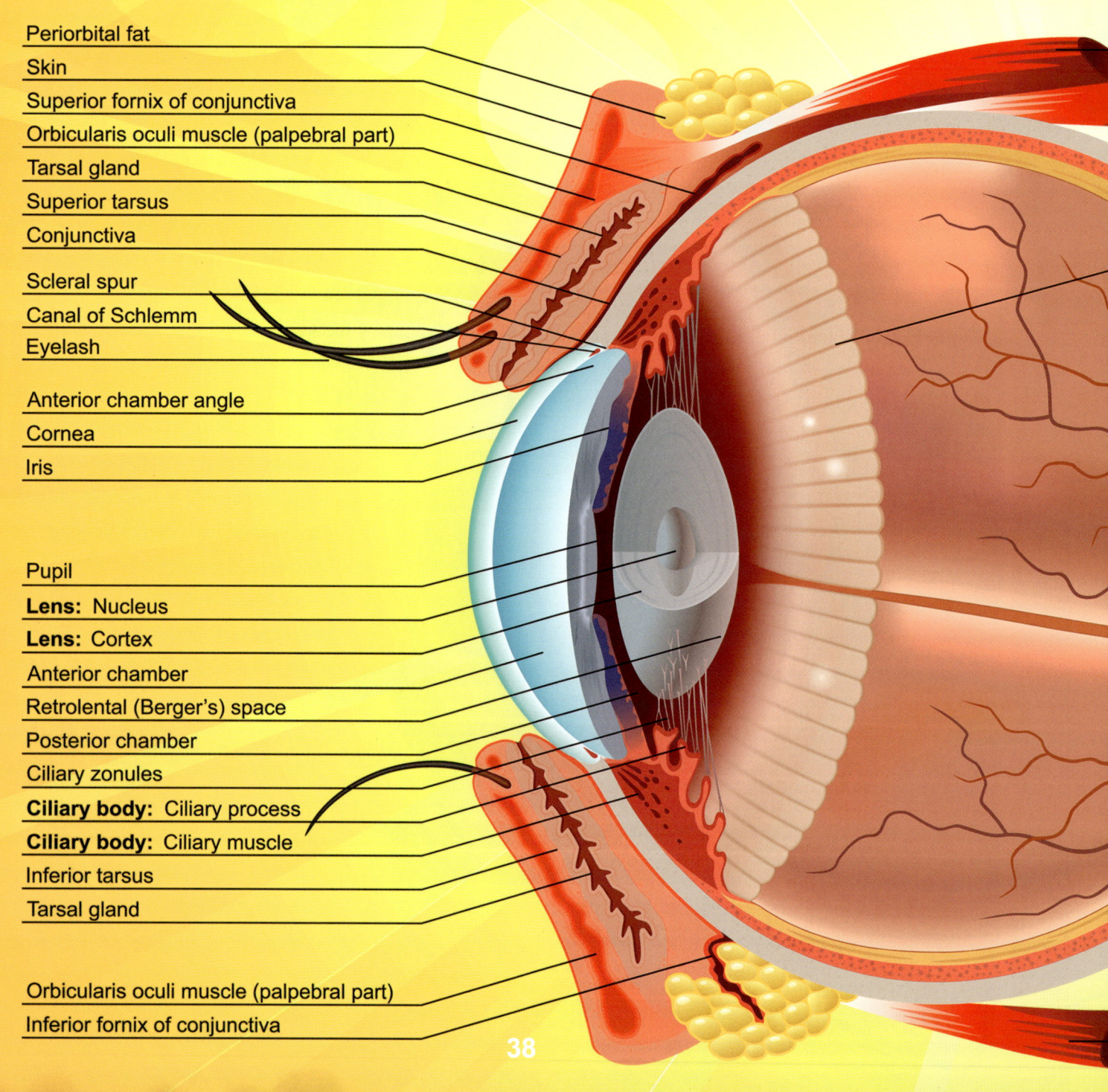

Periorbital fat
Skin
Superior fornix of conjunctiva
Orbicularis oculi muscle (palpebral part)
Tarsal gland
Superior tarsus
Conjunctiva

Scleral spur
Canal of Schlemm
Eyelash

Anterior chamber angle
Cornea
Iris

Pupil
**Lens:** Nucleus
**Lens:** Cortex
Anterior chamber
Retrolental (Berger's) space
Posterior chamber
Ciliary zonules
**Ciliary body:** Ciliary process
**Ciliary body:** Ciliary muscle
Inferior tarsus
Tarsal gland

Orbicularis oculi muscle (palpebral part)
Inferior fornix of conjunctiva

# The Blind Watchmaker

Evolutionists say that given enough time, a blind watchmaker can make a complex mechanical watch. The problem with this is that evolution is like a blind man, not a watchmaker. It has no previous experience or understanding about what it is going to make. Also, as complicated as a watch is, it is nothing compared to the complexity of living organisms.

The founder of evolution, Charles Darwin said it was absurd to imagine natural selection could produce the complicated lens of the eye. That was just the lens. What about all the rest of the eye. Remember, nothing works if anything is missing.

Now multiply that story a million or

muscle
muscle
serrata
Retina
Sclera

Choroid
Vitreous body

Fovea
Posterior ciliary arteries
Hyaloid canal
Lamina cribrosa
Dura mater
Arachnoid
Optic nerve
Pia mater
Central retinal artery
Central retinal vein
Optic disk

Vitreous body
Inferior rectus muscle

# THE LARGEST TREES ON EARTH
## can't survive by themselves

Taproots can be twice as deep as a tree's height. They go deep providing strength for the tree during storms. Does it surprise you to find out that the tallest trees in the world are redwood trees and they do not have taproots? they get much of their 500 gallons a day of needed water from absorbing fog through their bark. So, how do they stand without falling over during violent storms?

It turns out that they do it by teaming up with other large trees and intertwining their horizontal roots to form one large root system. They hold each other together with one giant mass of roots which spreads out over a large area like a foundation. The redwood tree is unique in the world in size and method of survival. Redwoods grow in clusters or groups. A single tree of that height could not stand against a violent storm by itself.

— Taproot —
like an anchor
to the earth.

These intertwined and
knotted roots are
stronger than a taproot.

The redwoods combine their
strengths to endure storms.

If the tallest redwood tree fell on a football field, it would be longer than the field.

Each redwood cone is about one inch long and produces 100,000 seeds the size of a tomato seed. Each seed is capable of growing a 300 foot tree!

1,000 year old redwood grove

Only redwoods have the strength and ability to support other redwoods.

# BACTERIA FLAGELLUM
## microorganism with an electric motor

It's tail (flagellum) is turned by an electric motor.

# RIDDLE

## Why is a Bacteria Flagellum like a MOUSETRAP?

(What does that mean? Keep reading)

**ANSWER:** Both are Irreducibly Complex!

**Explanation:** The mousetrap is constructed with five parts. There is a hammer, catch, spring, platform, and hold bar. If any one of those essential parts is missing, it will not work. It cannot be reduced; it cannot be less complex.

The bacteria flagellum motor has over 30 essential parts and if any one of them was missing, it would not work. There would not be bacteria, and actually, there would not be life on earth. If the first bacteria on the earth was not complete and all parts functioning perfectly, we would not be here. It could not happen gradually. When you add in a complex functioning DNA and all the other cell parts, there are thousands of pieces that cannot be missing. The possibility of this happening by blind accident is zero. This design required an amazing Creator.

The bacteria flagellum electrical motor can disassemble and reassemble itself while running. The motor stators change their number every thirty seconds, adapting to their environment. It does this while rotating as fast as 100,000 RPM!

Bacteria are the only organisms in the world with a motor.

Our bodies have 30 trillion bacteria in them. That is 30,000,000,000,000 propellers spinning in us all the time. No wonder we feel tired.

A bacteria is so small that if you lined up 25,000 of them, they would measure one inch.

Evolutionary scientists believe the bacteria was one of the first living organisms to evolve. How could the earliest life form be so complex?

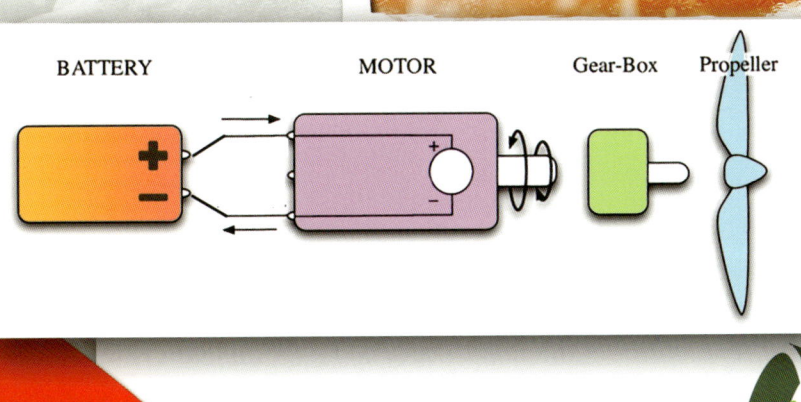

BATTERY      MOTOR      Gear-Box   Propeller

## IT WORKS LIKE AN ELECTRIC MOTOR !

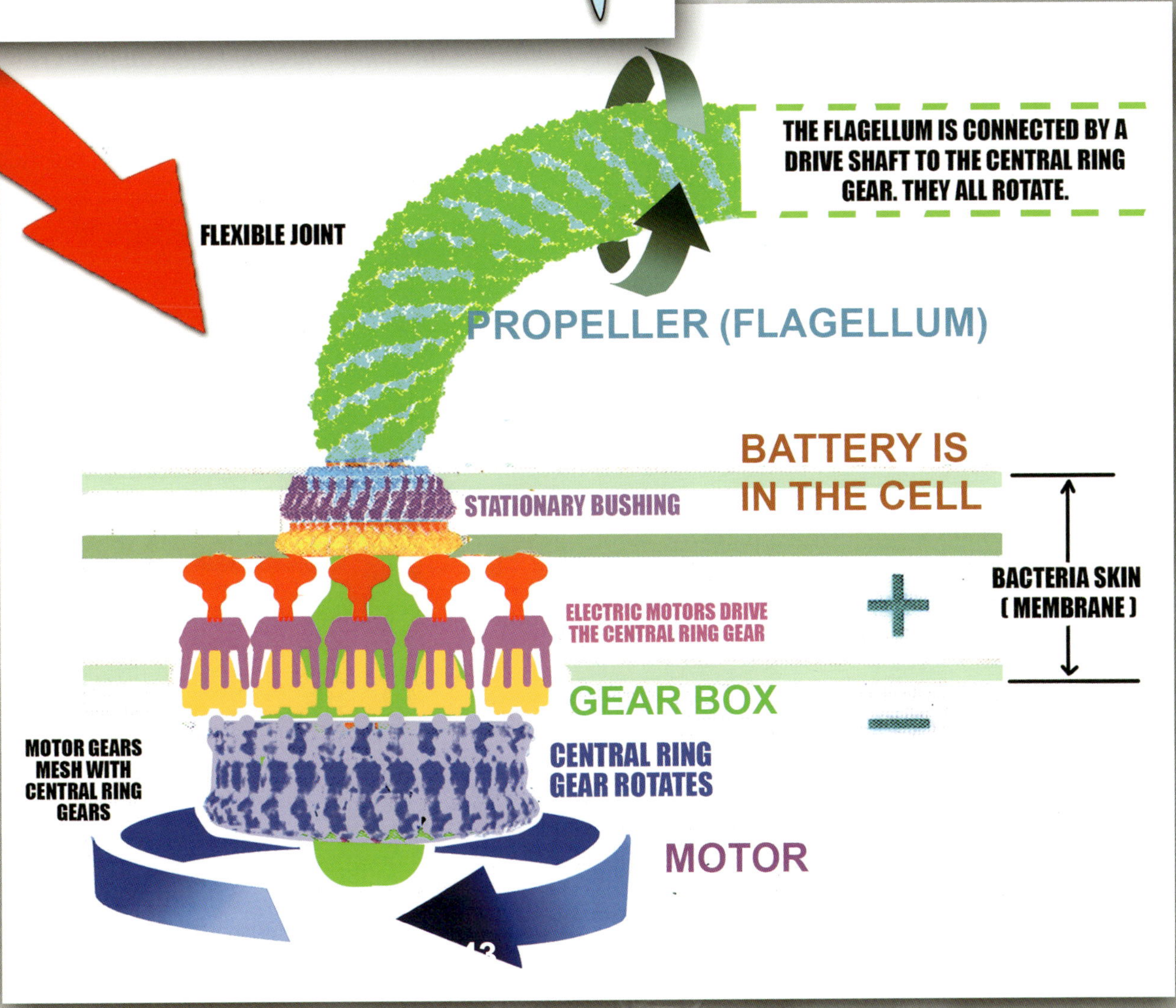

THE FLAGELLUM IS CONNECTED BY A DRIVE SHAFT TO THE CENTRAL RING GEAR. THEY ALL ROTATE.

FLEXIBLE JOINT

**PROPELLER (FLAGELLUM)**

**BATTERY IS IN THE CELL**

STATIONARY BUSHING

**BACTERIA SKIN (MEMBRANE)**

ELECTRIC MOTORS DRIVE THE CENTRAL RING GEAR

+

**GEAR BOX**

−

MOTOR GEARS MESH WITH CENTRAL RING GEARS

**CENTRAL RING GEAR ROTATES**

**MOTOR**

# THE FIREFLY
## THE ONLY INSECT WITH A TAIL LIGHT

There are about 2,000 different species of fireflies. Sometimes they are called lightning bugs, but they are neither flies nor bugs, they are beetles.

## HOW DO THEY DO THAT ??

Fireflies produce light in special organs in their abdomens. They formulate a chemical called luciferin and an enzyme called luciferases. When combined, these produce light. Fireflies control the flashing of the light by regulating the amount of oxygen that goes into the chemicals.

These wonderful beetles are also helping humans. The two chemicals produced in the firefly are being used in research on cancer, multiple sclerosis, cystic fibrosis and heart disease.

**THE ADULT FIREFLY ONLY LIVES ABOUT 3 WEEKS.**

44

## Telegraph Key

Before telephones, long distance communications were done by telegraph. Electric signals were sent on wires using dots and dashes, spelling words. It was called Morse Code. People could talk to each other all over the world.

### Blinking patterns of different species

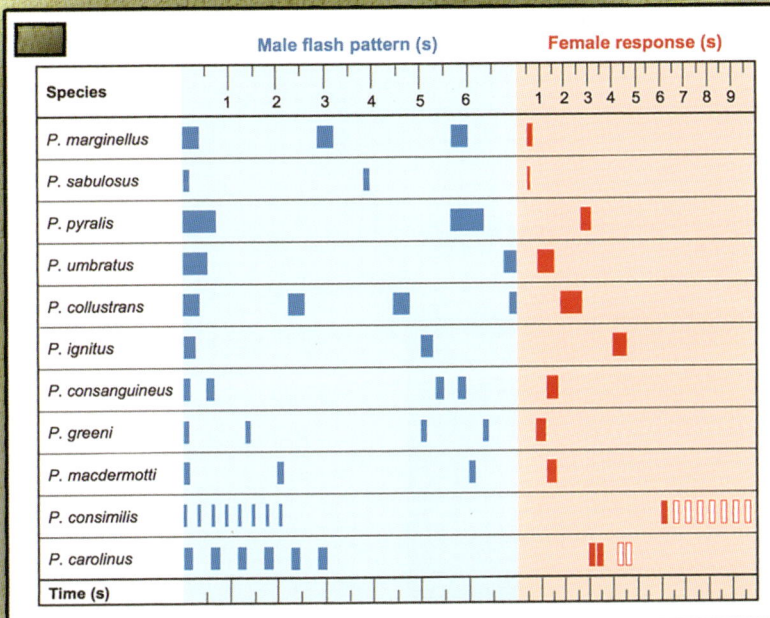

Fireflies use their lights as a form of Morse Code to talk to each other. They flash patterns of dots and dashes to identify themselves to other fireflies. Each species has a different pattern. Males use their flashing codes to find females or to warn of predators.

## FUN FACT

There is only one place in the USA where all the fireflies flash at the same time. They are like a beautiful choir singing to their Creator. Maybe they are singing, "This little light of mine, I'm gonna let it shine!"

GATLINBURG, TENNESSEE

# THE WEIRDEST AND LARGEST INSECTS
## the variety is endless

There are an estimated 10 quintillion insects in the world. That is

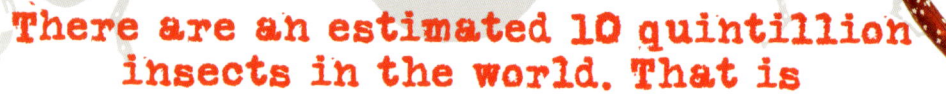

**10,000,000,000,000,000,000!**

**TRUNK BUTTERFLY**

**This is a small example of the millions of unique insect species on the earth.**

**BALL BEARING TREEHOPPER**

**PICASO BUG**

BY ALFRED KELLER

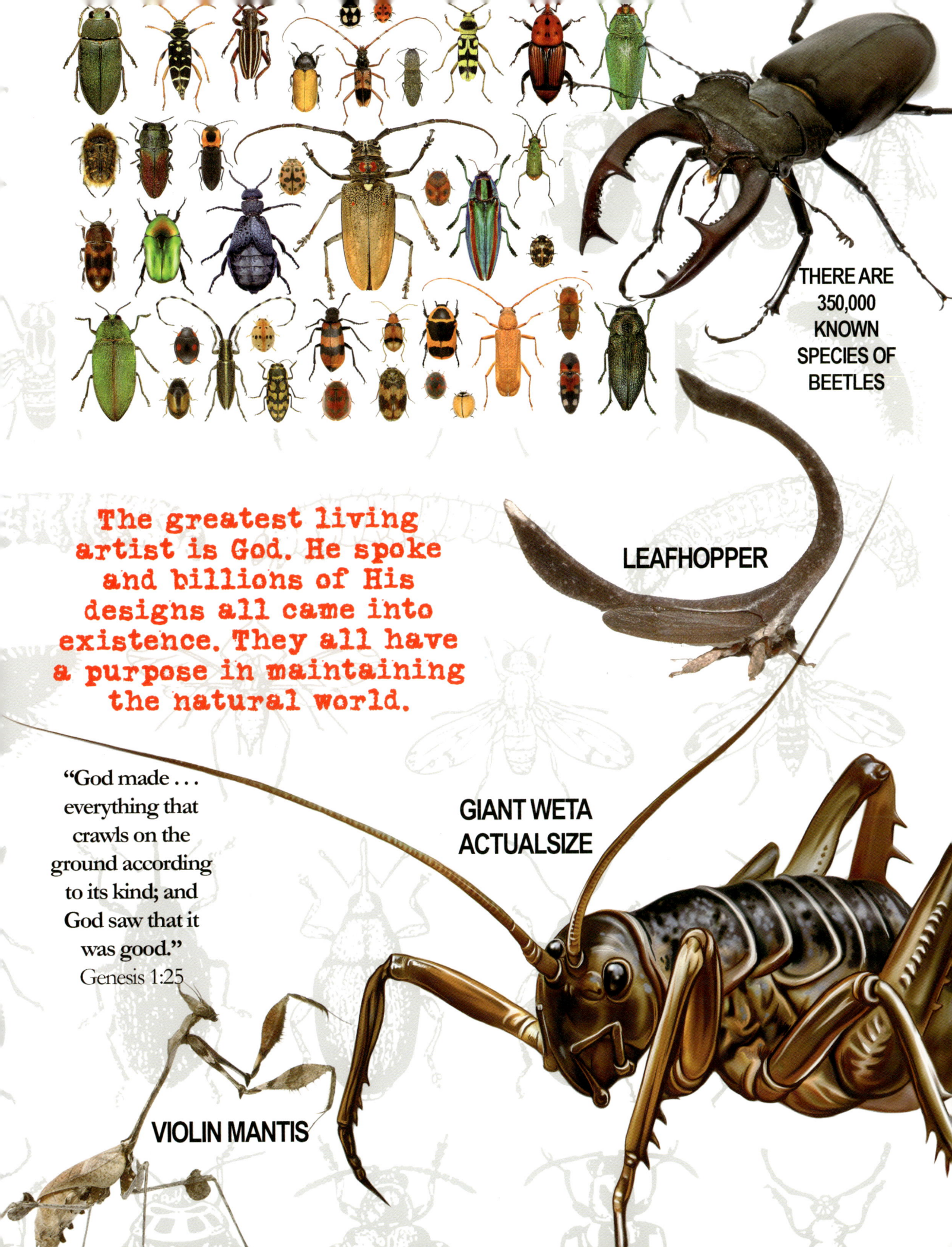

THERE ARE 350,000 KNOWN SPECIES OF BEETLES

The greatest living artist is God. He spoke and billions of His designs all came into existence. They all have a purpose in maintaining the natural world.

"God made . . . everything that crawls on the ground according to its kind; and God saw that it was good."
Genesis 1:25

LEAFHOPPER

GIANT WETA
ACTUALSIZE

VIOLIN MANTIS

# THE ARCHER FISH

## the fish that acts like a sniper

The archer fish can launch itself up to two feet out of the water.

It does this to grab prey before another fish can get to it first.

One more thing. They have even been seen jumping out of the water to catch flying insects.

Jaw muscles are used to pump water through a tube formed by its tongue and a unique channel in the roof of its mouth. The stream of water knocks its prey into the water where it is easily devoured.

THEY CAN HIT TARGETS 5 FEET ABOVE THE WATER.

WATERLINE

SPECIMEN TAG    Reference #154

Date:
Name: Archer Fish
Period:
Age:
Location: Swamps in Asia and Australia

# REFRACTION
## THE CHANGE IN DIRECTION OF A LIGHT WAVE PASSING FROM ONE MEDIUM TO ANOTHER.

When light passes from air to water, it is bent. Any object viewed will not be where it appears to be. An archer fish views its target insect from below the water so the insect is not where the fish sees it. The fish, however, calculates for the refraction to find the exact location of the insect. When it squirts its water, it rarely misses. It is a mystery to science how it can calculate this.

**REFRACTION**

Angle of incidence

Incident ray

90°

Boundary

Substance 1
Substance 2

## SURVIVAL OF THE FITTEST

*The natural process by which organisms best adjusted to their environment are most successful in surviving and reproducing.*

The archer fish does not need to jump or squirt water bullets to survive. It spends most of its time eating water bugs and spiders like other fish. It swims around and does fine without its amazing and unique skills. Science cannot explain why it can accurately knock bugs out of trees and jump out of the water and catch them since it does not need to do it. It seems to do it for sport, not for survival. Sometimes, two or three archer fish will work together. Our Creator made an incredible fish and science cannot explain why it does things it doesn't need to do.

TOXOTES JACULATOR

6 CENTS SINGAPORE

# THE AMAZING BAT
## they see with their ears!

**1** Bats make a series of loud clicks or pings creating sound waves.

Sound waves that bounce off the moth reach the ears of the bat.

**3**

Sound waves hit a flying insect and bounce back toward the bat.

**2**

## BAT SONAR AND HEARING

Bats produce "echolocation," (echo + location), or sonar, by emitting high-frequency sound pulses through their mouth or nose and then listening to the echo. The bat's hearing is so sensitive it can determine the size, shape, texture, and distance of objects around it. Bats can detect objects the size of a human hair. They are seeing with sound. The bat is unique. It is the only flying mammal with sophisticated sonar for its nighttime guidance.

Ships on lakes and the oceans, along with submarines use sonar to locate formations, fish, or underseas vessels that they cannot see from the surface of the water. They can all thank the bats for the technology. And while you are at it, thank the Creator of the universe for the idea.

Moth

# THERE ARE OVER 1,400 SPECIES OF BATS WORLDWIDE

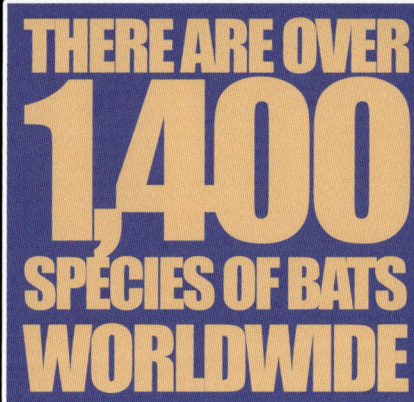

## Bats sleep Upside Down

# 25% of the world's mammals are Bats.

## 3% of the world's bats are vampire bats.

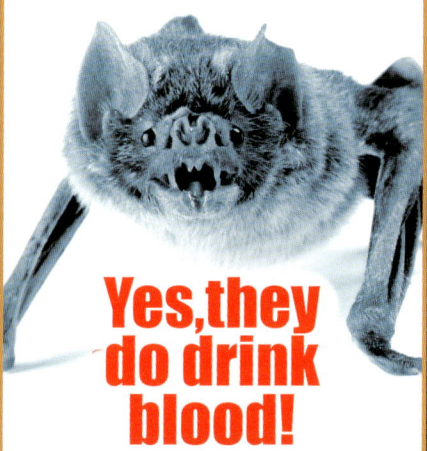

**Yes, they do drink blood!**

## DID YOU KNOW?

## Bats can eat 1,200 mosquitoes an hour.

## Bat's wings are like our arms and hands.

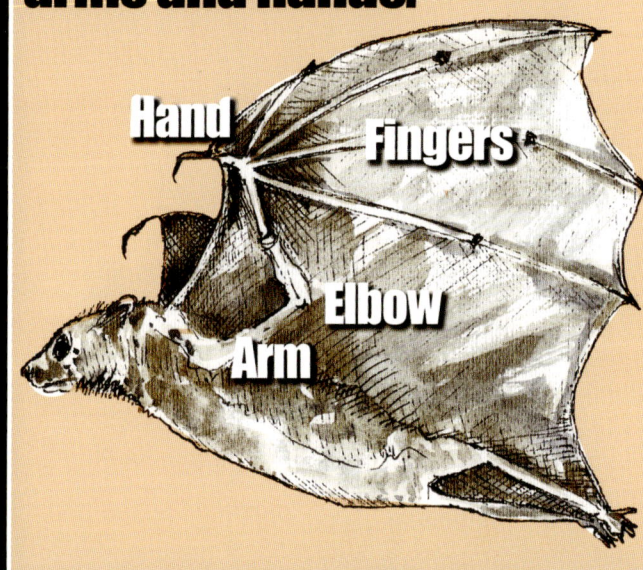

Hand
Fingers
Elbow
Arm

## The smallest mammal in the world is the Bumblebee bat from Thailand, which weighs less than a penny.

## THE LARGEST BATS ARE THE ASIAN FLYING FOXES.

72 Inches

Some people equate bats with death, graves, and Halloween. Yet, they are a needed part of the design of our natural world.

# BOMBARDIER BEETLE

## nature's explosive expert

This unique insect produces several caustic chemicals in its body and stores them in special chambers. When it is attacked it injects the chemicals into an explosion chamber lined with a special catalyst that makes the mixture explode in boiling fury upon the unsuspecting enemy.

**Species** : Bombardier Beetle
**Habitat** : Wooded areas
**Region** : Much of the world
**Diet** : Small Insects

SPECIAL

Danger
Caustic

## HOW DO THEY DO THAT ??

**Glands** that produce the two main chemicals that will produce a violent explosion, Hydrogen Peroxide and Hydroquinones

**Storage Chambers** for Hydrogen Peroxide Hydroquinones

**Glands** on the explosion chamber produce special catalyst enzymes that trigger the chemicals to explode

**Muscles** which control the flow and mixture of the powerful chemicals

**Explosion Chamber**

212 degree scalding caustic-steam explodes into the face of its attacker

RUFOUS BOMBARDIER BEETLE

STATE OF QATAR

Brachinus nobilis

POSTAGE 2 RIYALS

The beetle's spray nozzle can rotate in any direction like a turret on a military tank. The legs of the beetle are protected from the steaming spray.

No other creature produces hydroquinone and hydrogen peroxide in their body as a defense mechanism.

Instead of a continuous spray, the beetle uses machine gun style bursts of spray bullets at 500 explosions per second.

The pulses are necessary to stabilize the beetle, otherwise it would blow itself up or propel itself like a jet taking off.

## THINK ABOUT IT . . .

While most beetles fight their enemies with stinky smells or pinchers, this unique insect has a full chemical factory in its body. There is nothing else like it in the world. Some people think they evolved from other beetles. How could they keep from blowing themselves up while they were evolving? It never happened that way. God made this little bomb maker like He created millions of other unique creatures to fill the earth.

# THE BOWERBIRD
## builds a chapel instead of a nest

**"Bower" definition**
An area shaded by trees or other plants, a woman's private dressing room or a country cottage.

AUSTRALIA 35c

Regent Bower Bird

**They are found throughout Australia and New Guinea.**

# THE AMAZING BOWERBIRD STORY

## A TIME TO BUILD

The male bowerbird has an interesting way of attracting a mate; it builds an arched cathedral-type structure (bower) on the ground made of sticks and twigs to attract a mate. The bowerbird spends a week to two months building the structure.

## A TIME TO DECORATE

Once the structure is complete, the male bird decorates the bower surroundings with trinkets such as bottle caps, ballpoint pen lids, glass, plastics, cloth, and other items discarded by humans. Bowerbirds will also decorate with rocks, shells, feathers, bones, berries, and flowers to attract a mate.

## A TIME TO PAINT

Different species favor different colors. The striped gardener bowerbird prefers yellow, red, and blue objects. The fawn-breasted bowerbird favors green. Some males also "paint" their walls with a mixture of charcoal dust and saliva or plant juices. The bird uses his beak or a bit of chewed bark as a paintbrush. (San Diego Zoo Wildlife Alliance)

## A TIME TO DANCE

After the bower is complete with its decorations, it is ready for inspection by prospective mates. Upon the arrival of a female, the male bowerbird will dance for the female while holding one of its favorite trinkets in his beak. If the female approves, they will mate, and the female will fly away to build a nest for her eggs. The male will stay at his bower in hopes of attracting another female.

*"There is an appointed time for everything. And there is a time for every matter under heaven"*

Ecclesiastes 3:1

# THE HUMAN BRAIN

## world's greatest computer

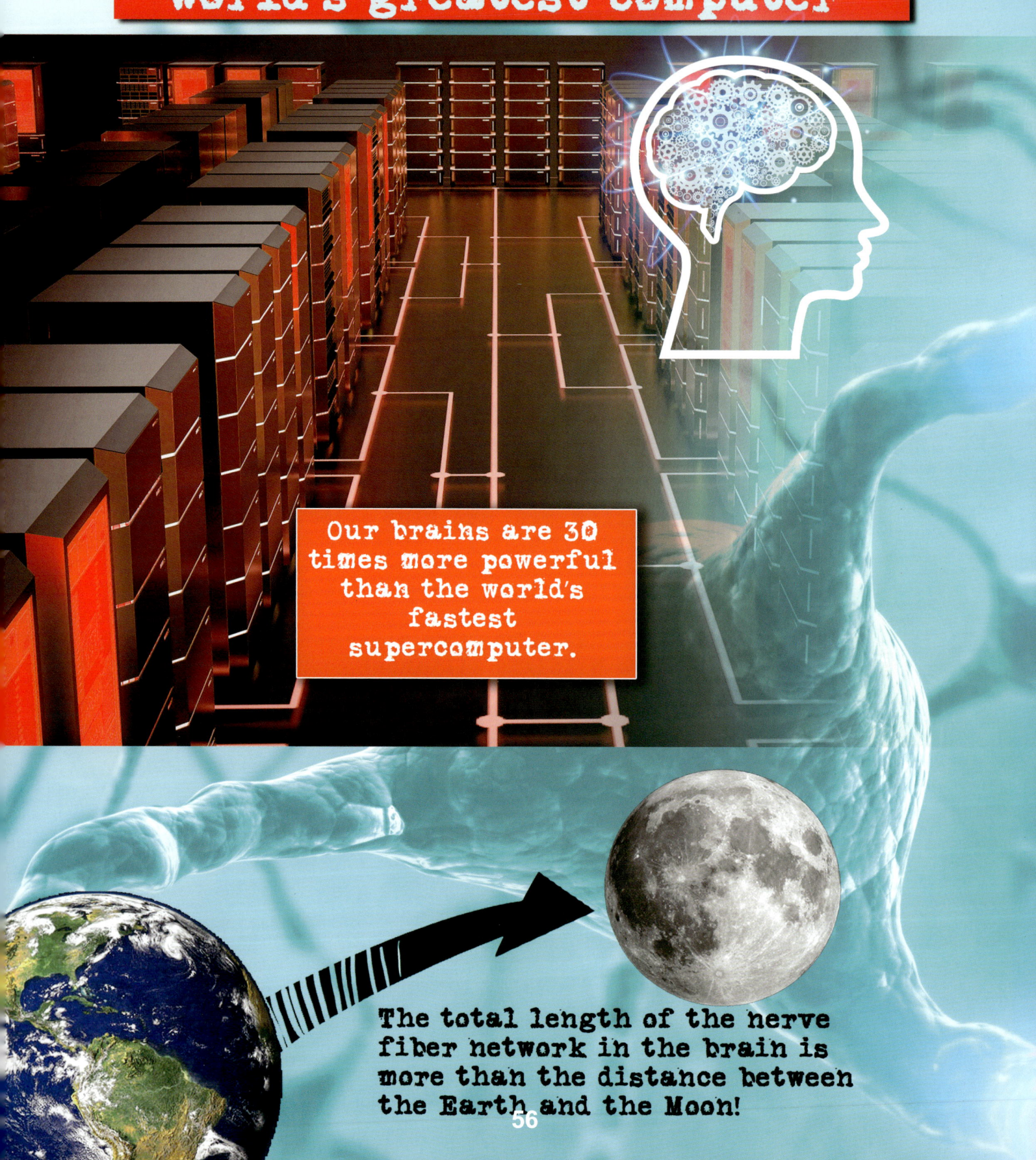

Our brains are 30 times more powerful than the world's fastest supercomputer.

The total length of the nerve fiber network in the brain is more than the distance between the Earth and the Moon!

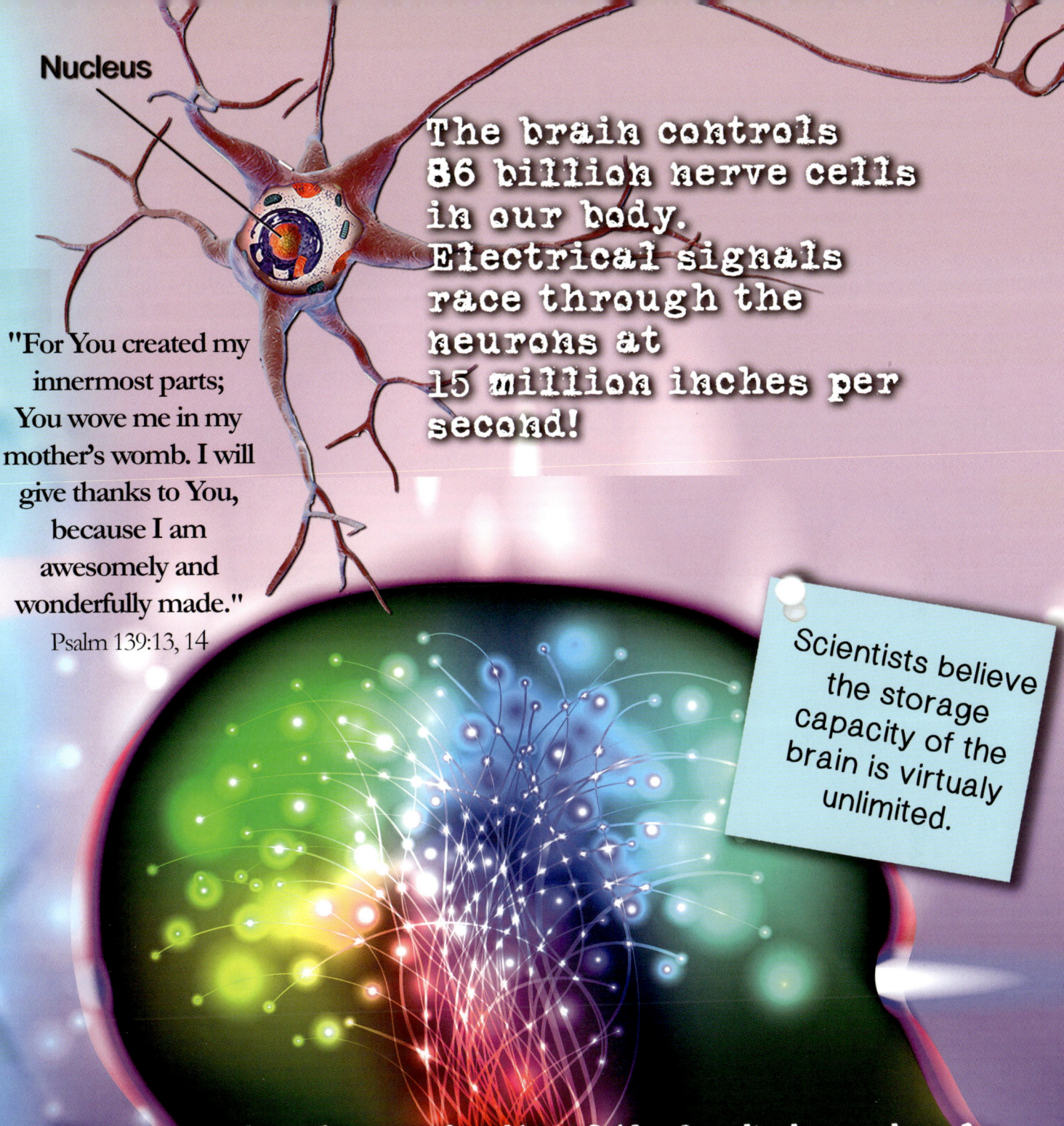

**Nucleus**

"For You created my innermost parts; You wove me in my mother's womb. I will give thanks to You, because I am awesomely and wonderfully made."

Psalm 139:13, 14

The brain controls 86 billion nerve cells in our body. Electrical signals race through the neurons at 15 million inches per second!

Scientists believe the storage capacity of the brain is virtualy unlimited.

The unimaginable complexity of the brain is a miracle of design by God. It is a mystery how the brain's electrical signals produce thoughts, dreams, consciousness, and memory?

Electricity flows through wires all over the world, but do electric wires dream, make artwork, write music, or think about the Creator that made them?

Use your brain!

# THE DUNG BEETLE
## nature's super duper pooper scooper!

## CLEANING UP THE MESS

Have you ever had to clean up dog poop? Gross! What if you had to pick up after an African elephant? Their mess can weigh 150 pounds. Yuck! Fortunately, there is a clean up expert, the dung beetle. They turn the poo piles into food for their families. They store it in underground caves. The caves are warehouses for fertilizer and seeds that grow new grass for the herds of animals in the region. The dung beetle is the only beetle that eats animal poop and fertilizes the earth.

## DWELLERS, TUNNELERS, AND ROLLERS

When a fresh pile of poop is located, many thousands of dung beetles fly to the site. Once there, they get to work. Yum! The TUNNELERS (diggers) carve out the caves. The DWELLERS set up the nests and organize the dung storage. The ROLLERS form the poop into balls and roll them to the caves. It is a large construction site with workers doing their different jobs. When the clean up job is complete, the crew flies off to the next fresh poop pile. The eggs left behind will hatch and become the next clean up crew. Yay! The seeds left behind in the ground will grow into new plants to feed the animals. Some people think it all happened by accident. A better explanation is that this was designed for a specific task by the Creator.

POOP HAPPENS

JUST PICK IT UP AND MOVE ON

Thank You

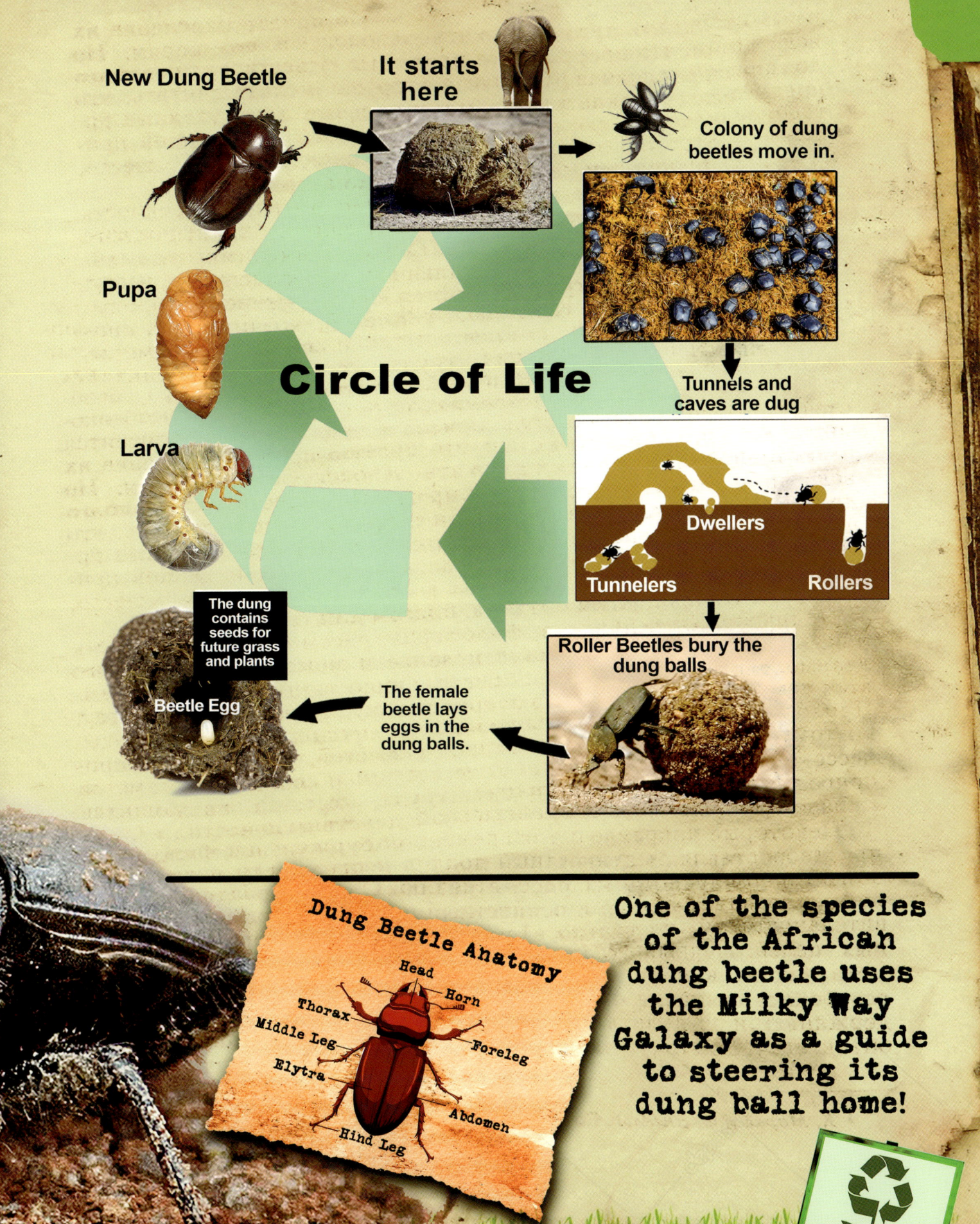

**New Dung Beetle**

**It starts here**

Colony of dung beetles move in.

**Pupa**

# Circle of Life

Tunnels and caves are dug

**Larva**

Dwellers

Tunnelers          Rollers

The dung contains seeds for future grass and plants

Roller Beetles bury the dung balls

**Beetle Egg**

The female beetle lays eggs in the dung balls.

Dung Beetle Anatomy

Head

Horn

Thorax

Foreleg

Middle Leg

Elytra

Abdomen

Hind Leg

One of the species of the African dung beetle uses the Milky Way Galaxy as a guide to steering its dung ball home!

# THE CROCODILE
## and his dentist

The Nile crocodile gets the award for being the most dangerous reptile. It is responsible for more than 300 fatal attacks on people each year.

They are feared by every animal. Their strong jaws clamp down harder than any other creature and they don't let go. They can actually bite through steel.

There is one creature that does not fear it. It is actually a friend and welcomed by the crocodile. It is a small bird. The bird needs the crocodile and the crocodile needs the bird. It is an Egyptian plover. It is a most unusual relationship.

*"But just ask the animals, and have them teach you; And the birds of the sky, and have them tell you. . .*

*". . . the hand of the Lord has done this,"* (Job 12:7, 9)

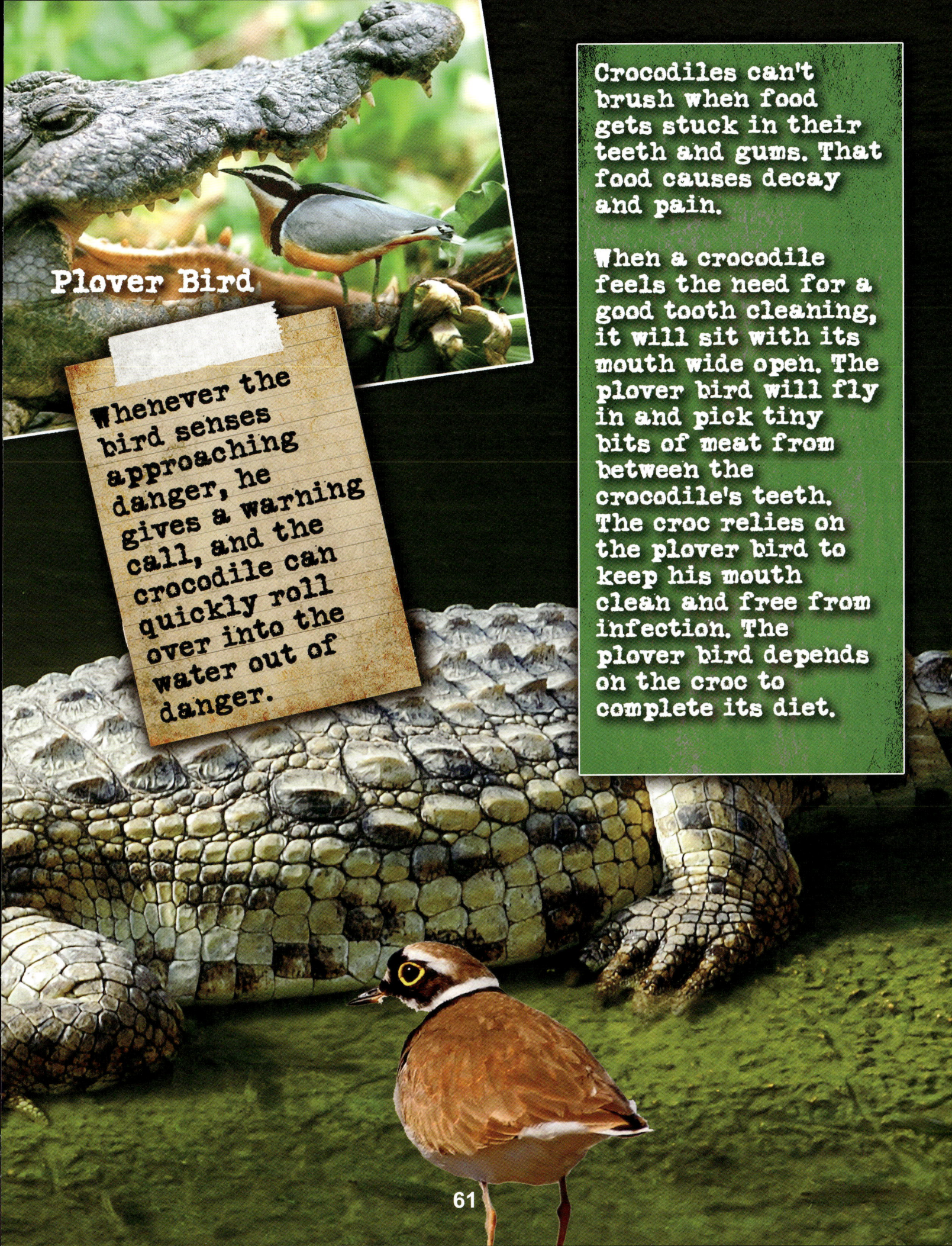

**Plover Bird**

Whenever the bird senses approaching danger, he gives a warning call, and the crocodile can quickly roll over into the water out of danger.

Crocodiles can't brush when food gets stuck in their teeth and gums. That food causes decay and pain.

When a crocodile feels the need for a good tooth cleaning, it will sit with its mouth wide open. The plover bird will fly in and pick tiny bits of meat from between the crocodile's teeth. The croc relies on the plover bird to keep his mouth clean and free from infection. The plover bird depends on the croc to complete its diet.

# EARTHWORMS
## "the intestines of the earth"
### Aristotle – 360 B.C.

Imagine a magic tube that you could put garden waste in one end, and when you push it through, nutritious soil comes out the other end. You just imagined an earthworm.

You might think, "What could be more simple than a worm?" They look like gooey pond scum or a gummy worm candy.

But when you look inside, they are amazingly complex. They even have five hearts!

Earthworms increase soil aeration, filtration, structure, nutrient cycling, water movement, and plant growth. Life as we know it would not exist without earthworms.

Earthworm castings (poop) are rich in iron, sulfur, calcium, nitrogen, phosphorus and potassium

**FORMULA FOR THE BEST SOIL CONDITIONER ON THE EARTH**

COMPOST

**+**

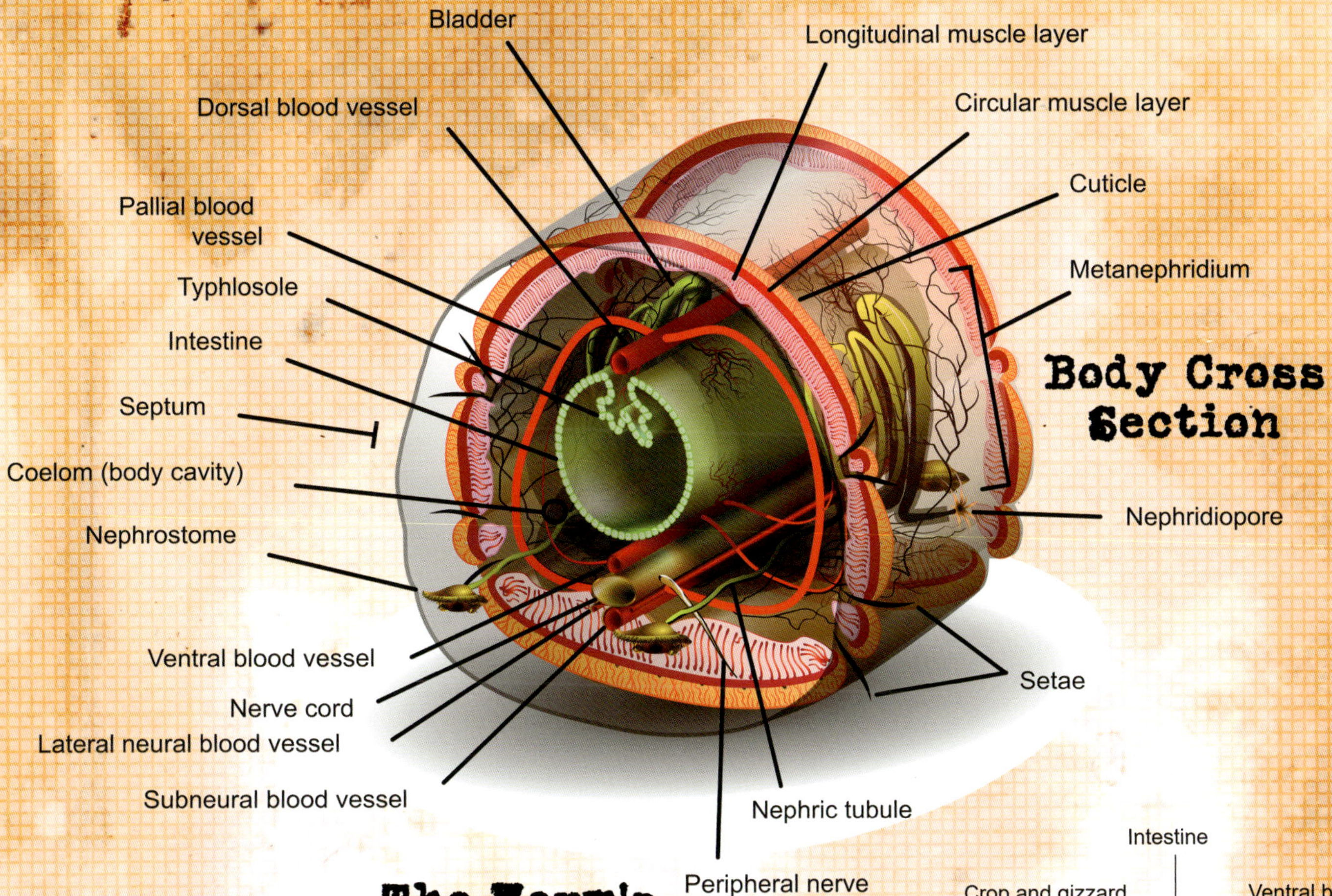

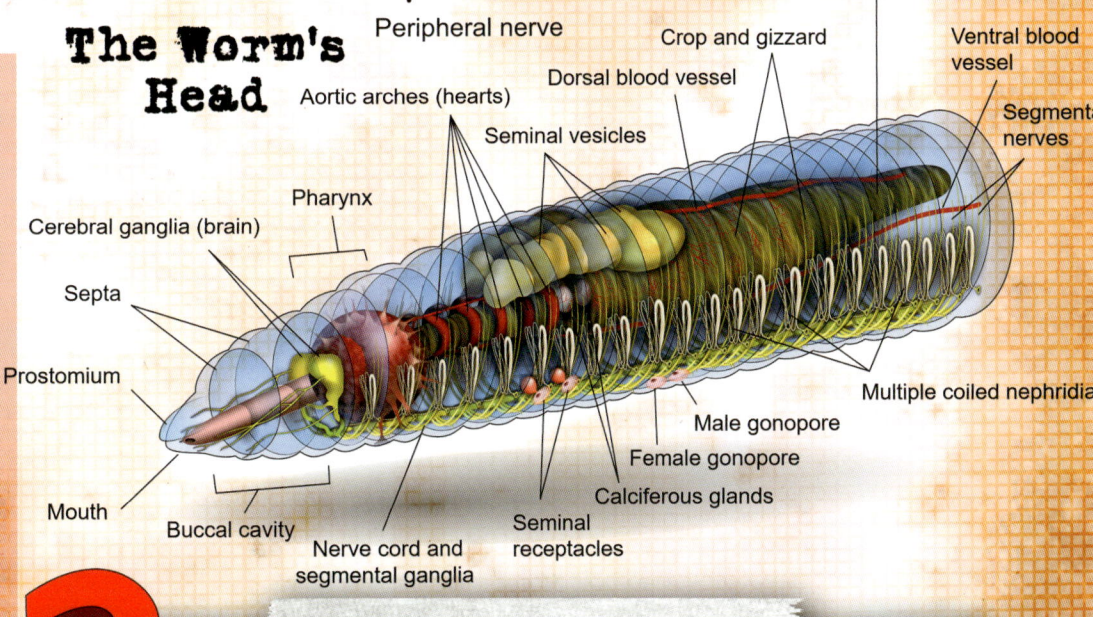

Bladder

Dorsal blood vessel

Longitudinal muscle layer

Circular muscle layer

Cuticle

Metanephridium

Pallial blood vessel

Typhlosole

Intestine

Septum

Coelom (body cavity)

Nephrostome

## Body Cross Section

Nephridiopore

Ventral blood vessel

Nerve cord

Lateral neural blood vessel

Subneural blood vessel

Setae

Nephric tubule

Peripheral nerve

## The Worm's Head

Aortic arches (hearts)

Pharynx

Cerebral ganglia (brain)

Septa

Prostomium

Mouth

Buccal cavity

Nerve cord and segmental ganglia

Seminal vesicles

Dorsal blood vessel

Crop and gizzard

Intestine

Ventral blood vessel

Segmental nerves

Multiple coiled nephridia

Male gonopore

Female gonopore

Calciferous glands

Seminal receptacles

Yes, the simple earthworm is extremely complex. If you remove any of its vital organs it would die. It has five hearts, a brain, DNA, a stomach, a digestive system, blood vessels, a central nervous system and more.

How did the highly complex earthworm come into existence? It was either created by God to generate soil and enrich plants, or it somehow evolved from an explosion which filled the universe with dead rocks, and by some unknown process produced life. What do you think?

# THE FRILLED LIZARD
## dragon with an umbrella

At first glance they look like any other lizard. But their defense mechanism is unlike any other lizard.

**FILM** **FILM**

The frilled lizard may look familiar to you if you saw the movie, "Jurassic Park." It was the model for their venom-spitting lizard, the Dilophosaurus.

7 7A

### Did You KNOW?

Frilled lizards are members of the dragon lizard family that live in the tropical and warm temperate forests and woodlands of Northern Territory of Australia. They spend most of their lives in the trees and feed on ants and small lizards.

The pleated and folded skin along with the collapsible bone and cartilage structure forms the frill framework. It is an example of engineering and design. The lizard can quickly expand and contract its large frill in the blink of an eye.

Most lizards run from predators. This one scares them off. If that doesn't work they retreat in a very unusual upright posture with their legs swinging outward which confuses the predator.

The uniqueness and beauty of this creature have the fingerprints of a Master Engineer and Designer.

## DEFENSE STRATEGY
1. Lay still.
2. Open frill.
3. Shriek and stare.
4. Run away upright.

# JELLYFISH
## underwater ballet

Fine art requires a great artist. Did Michelangelo's David just happen from a slab of marble rolling down a mountain? Did DaVinci's Mona Lisa result from a paint spill? When you look at the beauty of the hundreds of jellyfish in the sea and the varieties of design and function, it only adds to the overall evidence that a Master Designer is at work.

*"They, too, observed the LORD's power in action, his impressive works on the deepest seas."* (NLT, Psalm 107:24)

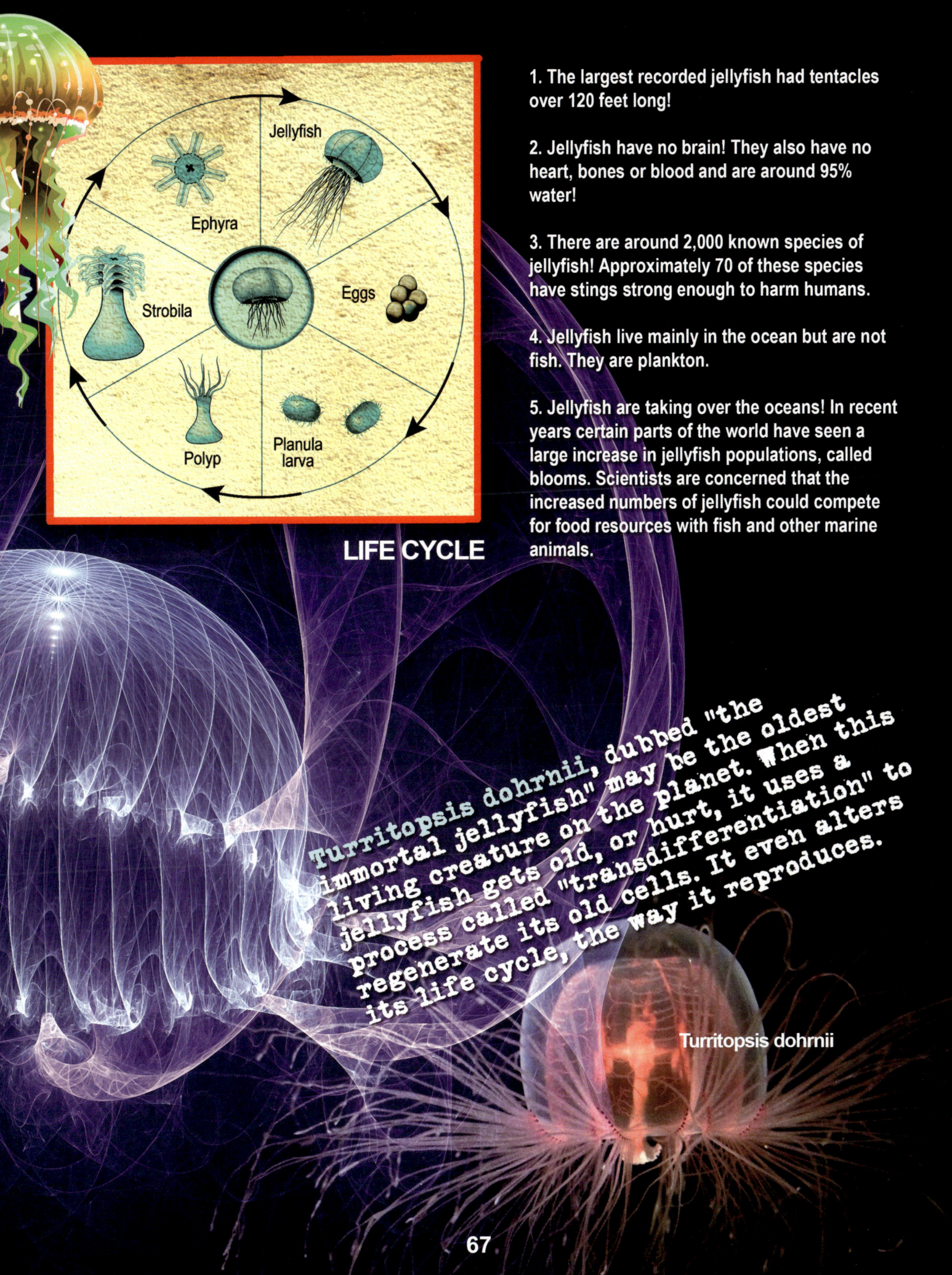

**LIFE CYCLE**

(Life cycle diagram labels: Jellyfish, Ephyra, Strobila, Polyp, Planula larva, Eggs)

1. The largest recorded jellyfish had tentacles over 120 feet long!

2. Jellyfish have no brain! They also have no heart, bones or blood and are around 95% water!

3. There are around 2,000 known species of jellyfish! Approximately 70 of these species have stings strong enough to harm humans.

4. Jellyfish live mainly in the ocean but are not fish. They are plankton.

5. Jellyfish are taking over the oceans! In recent years certain parts of the world have seen a large increase in jellyfish populations, called blooms. Scientists are concerned that the increased numbers of jellyfish could compete for food resources with fish and other marine animals.

Turritopsis dohrnii, dubbed "the immortal jellyfish" may be the oldest living creature on the planet. When this jellyfish gets old, or hurt, it uses a process called "transdifferentiation" to regenerate its old cells. It even alters its life cycle, the way it reproduces.

Turritopsis dohrnii

# LEAF ME ALONE!
## insects in disguise

The front

The back of the same butterfly

When threatened it mimics a falling leaf, floating to the ground.

with wings folded up

## DEAD LEAF BUTTERFLY

When the wings are closed, this species looks exactly like a dried autumn leaf, giving it the cleverest camouflage a butterfly could want. But when the wings are open, it reveals a luminous color pattern that can hold its own against the world's prettiest wings.

This incredible butterfly changes its colors with the seasons, just like a leaf does.

Dead leaf grasshopper

Leafy sea dragon

Walking stick leaf bug

The orchid praying mantis

Another mantis

The buff tip moth may be the best camouflaged insect in the world. If you didn't see the little legs, you would think it was a twig.

When it unfolds its wings, it looks like a moth.

There are thousands of bugs and creatures that look like leaves and branches. The possibility of these evolving from a rock, or from nothing, is impossible. They are the work of a Great Artist, the One who created all things.

# THE LEAFHOPPER
## insects that grow gears

Man-made gears

Leafhopper gears

Actual size

**DID YOU KNOW?**

The leafhopper's gears are the only functioning mechanical gears to be discovered in a living organism.

Its tiny teeth are located on the inner surface of each hind leg. The gear cogs remain engaged and roll past each other at a rate of nearly 50,000 teeth per second!

A leafhopper can jump 100 times their length. If a person could leap like the leafhopper, we could jump over the length of two football fields.

Look at an old mechanical watch with all its gears and springs. Where did it come from? Did the parts assemble themselves? It had to have a designer and skilled craftsman to make it. If any piece is missing, or slightly out of alignment, the watch will not work.

## Evolution's Nightmare

Since no other insect has gears, where did they come from?

The tiny, perfectly formed gears are finer than any thing man can make and they are really small. If you placed the gears next to each other in a row, it would take 1,000 of them to go across a single human hair! It takes the most powerful microscope to even see them. Evolutionists think this happened by accident.

If any part of the gears did not form perfectly they would be useless. They could not have taken millions of years to slowly evolve. Until they were perfect, they could not be used. The laws of evolution would have eliminated them.

The unique gears in the leafhopper had to appear all at once, perfectly formed, with all the muscles and nerves to make them work.

The only way this could have happened is that God created this insect!

# THE GIRAFFE
## changes its blood pressure

God gave the giraffe an amazing blood pressure regulation system. Blood vessels are controlled by the giraffe. Every valve, muscle, and bypass in the system have to be in place or the animal will die. If one part is missing – no giraffe.

Evolutionists believe the Okapi changed and stretched over millions of years and became a giraffe. Why do we see both of them today and nothing in between?

Giraffe

Okapi

*"Then God said, "Let the earth produce living creatures according to their kind: livestock and crawling things and animals of the earth according to their kind"; and it was so."*
Genesis 1:24

A giraffe's heart is 2 feet long and weighs about 25 pounds.

The height of the giraffe makes it hard for the heart to pump blood to the brain. A series of one-way valves force blood toward the head.

When the giraffe is drinking, its blood vessels constrict (shrink) to lower its blood pressure so its brain is not damaged.

A giraffe's neck is too short to reach the ground.

1. Unique pressure sensors on the arteries detect change in blood pressure.

2. Strong neck muscle strands in the arteries constrict blood flow which lowers blood pressure.

3. A series of valves in the veins control the blood return to the heart.

4. Some arteries near the brain are able to bypass blood back to the heart avoiding the brain.

The entire system of muscles, nerves, valves, and bypasses keep the blood pressure controlled whether the giraffe is standing or kneeling.

# THE CELL MITOCHONDRIA

## complex power grid

Mitochondria make electrical power to run the cells in our body. Like any power plant they wear out. There are about 10 million billion mitochondria in an adult human! Two billion mitochondria are made every second throughout a person's life. The lifespan of a single mitochondrion averages around 100 days. Even more amazng is how complex they are.

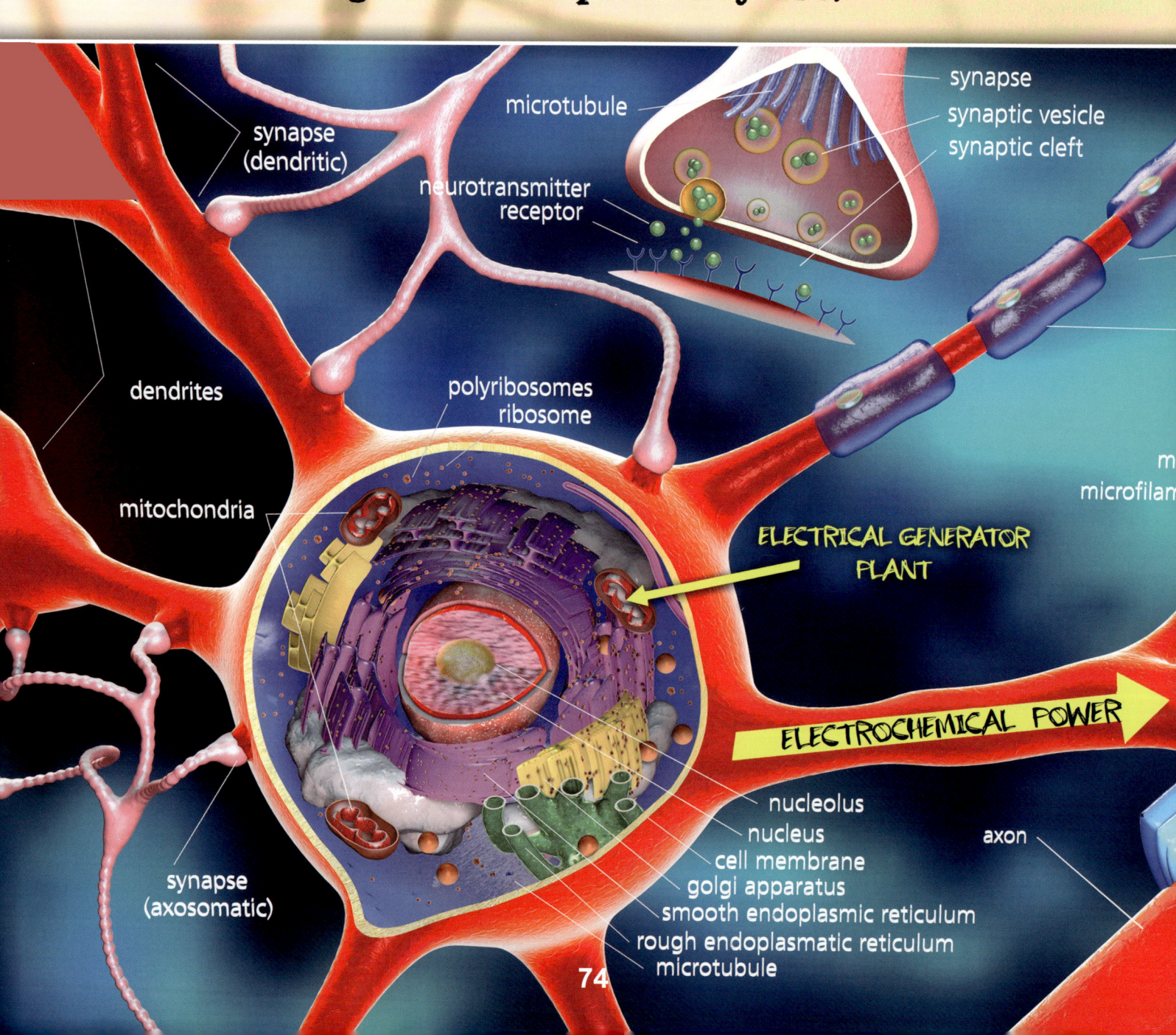

synapse (dendritic)

microtubule

synapse
synaptic vesicle
synaptic cleft

neurotransmitter receptor

dendrites

polyribosomes
ribosome

mitochondria

ELECTRICAL GENERATOR PLANT

m
microfilam

ELECTROCHEMICAL POWER

nucleolus
nucleus
cell membrane
golgi apparatus
smooth endoplasmic reticulum
rough endoplasmatic reticulum
microtubule

axon

synapse (axosomatic)

74

# A Nerve Cell

**Nucleus**

There are 86 billion nerve cells in our body.

**THE ENTIRE LENGTH OF A NERVE CELL IS LESS THAN THE WIDTH OF A HUMAN HAIR**

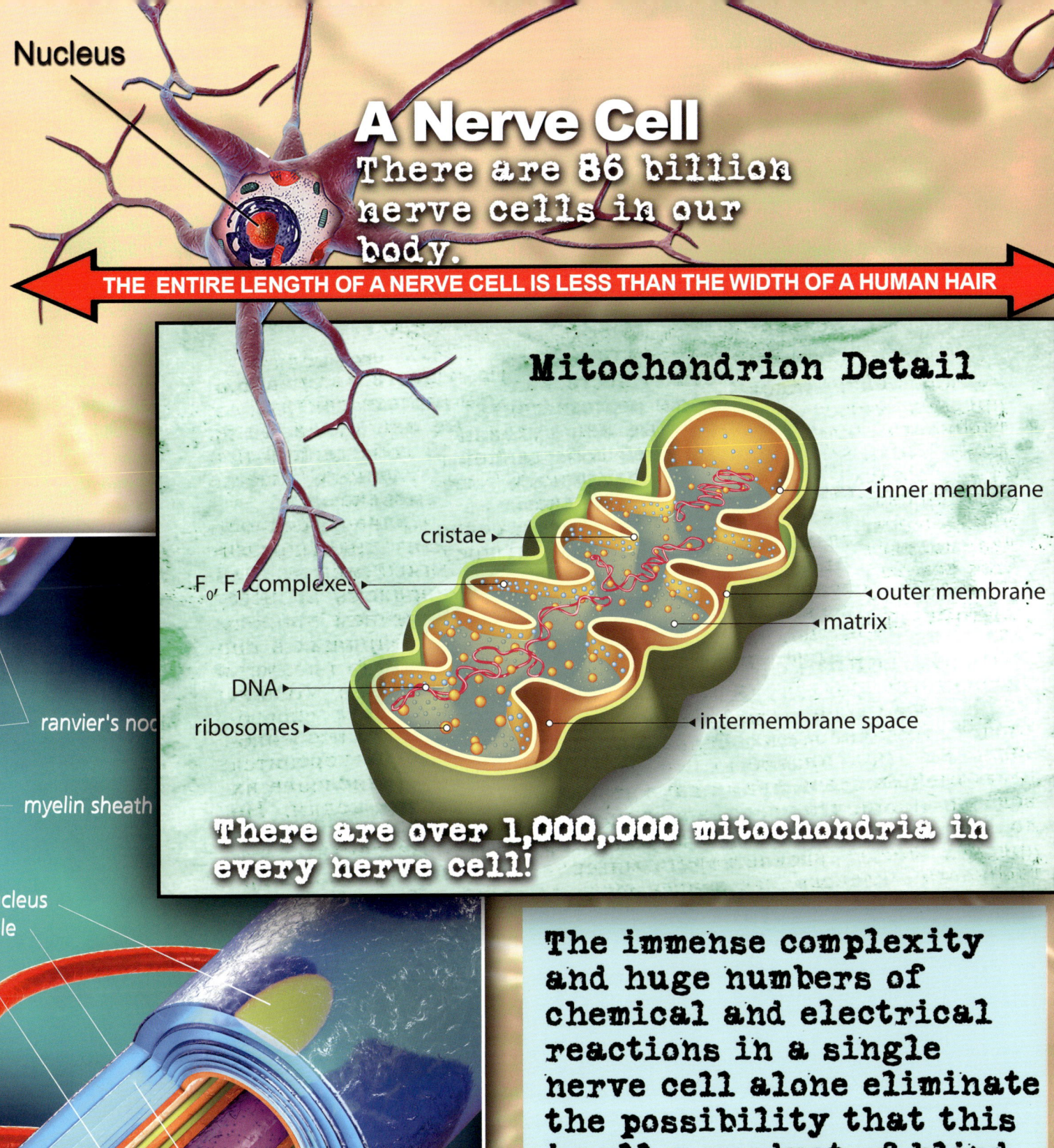

## Mitochondrion Detail

- cristae
- $F_0$, $F_1$ complexes
- DNA
- ribosomes
- inner membrane
- outer membrane
- matrix
- intermembrane space

**There are over 1,000,000 mitochondria in every nerve cell!**

ranvier's node

myelin sheath

nucleus
bule

The immense complexity and huge numbers of chemical and electrical reactions in a single nerve cell alone eliminate the possibility that this is all a product of blind chance. Remember, these cell power plants are only one of many complex operations in a living cell. Each function is needed for it all to work.

# THE MOSQUITO
## deadliest creature on earth

## THE FACTS

Diseases that are spread to people by mosquitoes include Zika Virus, West Nile Virus, Chikungunya Virus, Dengue, Yellow Fever, Encephalitis, Rift Valley Fever, and Malaria. Mosquitoes kill an estimated 750,000 to 1 million humans yearly.

### The Proboscis

The proboscis is like a hollow tube. A mosquito bite is like getting an injection with a needle, except the mosquito has six needles in the tube. Two of them are jagged, like saws to cut through our tough skin. The others all move independently and locate a blood vessel to suck out blood. Only the female does this to aid them in producing eggs.

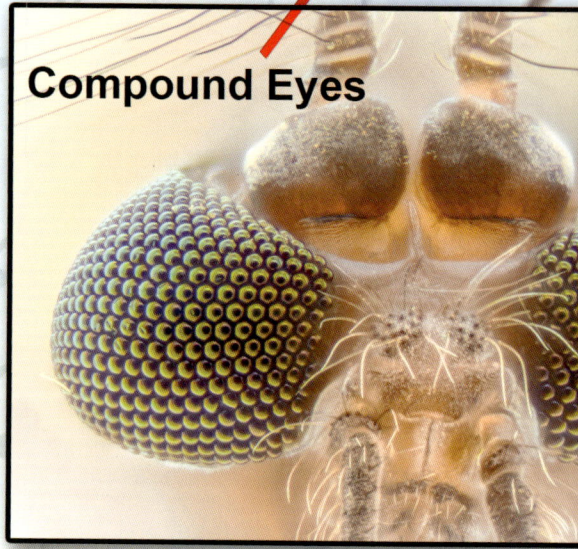

**Compound Eyes**

Mosquitoes have two compound eyes located on the sides of their heads. The eyes are covered with thousands of specialized lenses and each lens functions as an individual eye. They allow the mosquito to see in multiple directions. It is one of the most complex eyes in all of creation.

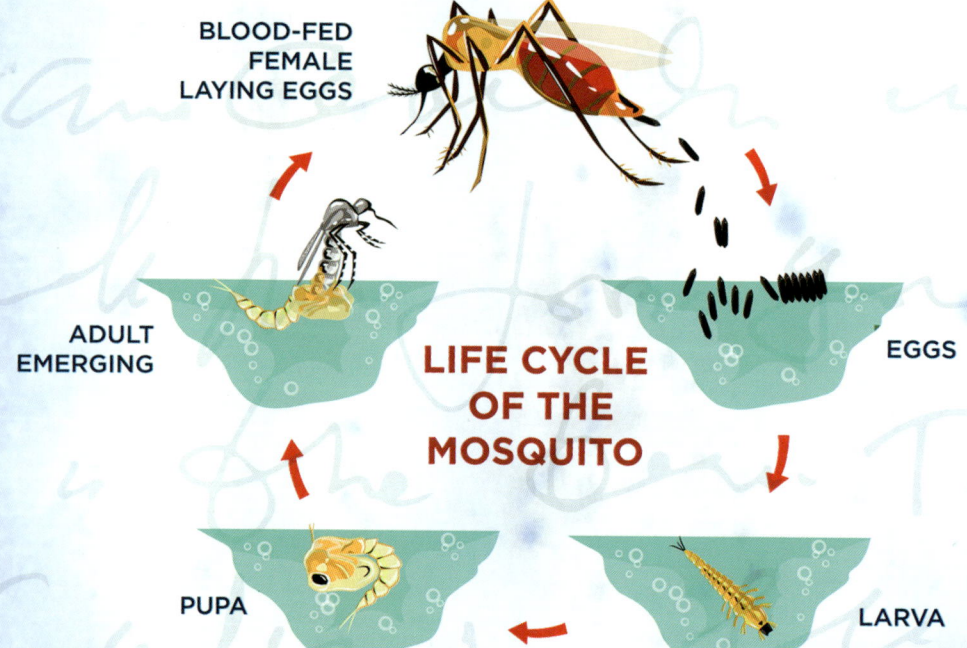

LIFE CYCLE OF THE MOSQUITO

BLOOD-FED FEMALE LAYING EGGS

ADULT EMERGING

EGGS

PUPA

LARVA

Here are some fun facts

- There are more than 3,000 species. Only a couple hundred feed on human blood.
- Malaria infects around 250 million people each year worldwide and kills about one million, mostly children in Africa.
- Only female mosquitoes bite.
- Mosquitoes are one of the slowest insects, 1 to 1.5 miles per hour.
- Mosquitoes carry heart worm which can seriously affect the health of your pets and even kill dogs.
- That buzzing sound in your ear is a result of rapid wing beating. A mosquito beats its wings 300 to 600 times per second.
- The average mosquito lives less than two months.
- Mosquitoes are a source of food for birds, bats, and frogs.

## Why did God create mosquitoes?

"God created mosquitoes as part of His intricate design for a perfect world. Unfortunately, when mankind, God's divinely appointed caretakers of the world, rebelled against their Creator, they plunged the whole world into a state of misery and decay. The discomfort and disease caused by mosquitoes are part of the groaning that all creation experiences (Romans 8:22). The good news is that Jesus is coming back to liberate creation from its bondage and reveal the glory of the children of God (verses 18–21). Jesus will fix our broken planet. There will be no more hunger, disease, or blood-sucking mosquitoes."

- Got Questions Commentary

# THE MALLEEFOWL
## the thermometer bird

Birds have nests, big deal, right? This book is about unique creatures and things they do. The Malleefowl makes the most unique nest in the world. It has a perfect climate control system that keeps the nest at exactly 92 degrees in summer, winter, all seasons. The secret is the male Malleefowl. He continually samples the soil with his beak and tongue which is a highly accurate temperature probe. Keep reading to see what he does next.

The Malleefowl from Austalia holds the world record for the largest nest.

It was measured at 15 feet in height and 35 feet across. To excavate and build it, the Malleefowl moved 8,793 cubic feet of material weighing 300 tons (661,386pounds). That would fill eight 20' shipping containers. One bird did that.

X 8

NEST

MALLEE FOWL

The female lays her eggs after the huge nest is built.

**3 Waits for rain**

The rain activates the compost causing it to produce heat.

**2 Builds compost pile**

3 feet deep

**1 Bird digs large hole**

92°F

**6 The nest is excavated, eggs are laid, then covered with sand.**

**5 Then, the mound is constructed**

92 degrees

**4 Compost pile is covered with sand**

Sand

Rotting compost produces heat.

How the Malleefowl does it.

**7**

The Malleefowl sticks its beak into the sand continually and can read the mound temperature like a thermometer. He then moves and removes dirt to add or subtract insulation, thus maintaining the 92 degree temperature needed to hatch the eggs. Once the chicks are born they need to dig their way out and within 24 hours they fly away without any help from their parents. No other bird does this. There are too many unique things to call this an accident. The Mallee is another example of God's creative design.

I may be the size of a chicken, but I work harder!

# OCTOPUS, SQUID, AND CUTTLEFISH
## masters of disguise

**CUTTLEFISH**

**SQUID**

### How do they change their color and shape?
#### (Cephalopods)

Have you ever squished a water balloon with your hand? It bulged out and became smaller when you stopped. The multilayer skin of these unique creatures has millions of vessels, like tiny balloons, each filled with a different color dye. Each balloon has its own muscles and nerves expanding or shrinking to change their skin into an endless variety of colors. They look like an underwater high-def TV.

The sacks of color are called
#### Chromatophores

They are unique; no other creature in the sea has this skin or ability. They can also control the muscles in their skin to resemble the rocks on the ocean floor. This happens in less time than you can blink your eye. This is not just amazing, it is miraculous. The immensity of the operation cannot be explained as simply happening by accident. It is a highly complicated and unique design created by God.

**Octopus Skin**

Shape - Shifting Champion

The Cuttlefish

The same Cuttlefish

These amazing creatures change their skin's color, brightness, contrast and pattern in as little as one-fifth of one second, as fast as the blink of an eye!

The octopus' arm is covered in 2,240 suction cups that are used to grip, taste, and smell. Each suction cup has more taste buds than the human tongue.

The octopus can survive 20-30 minutes outside of the water and have been known to roam the shores at night in search of food. The 800 living species of cephalopods are considered to be highly intelligent.

The largest cuttlefish grows to around 20 inches.

Cephalopods squirt ink to confuse predators. It is like the counter-measures submarines use to confuse the torpedoes of enemy subs.

# THE DREADED WASP
## a really good guy

**Interesting**

God used hornets to make Joshua's enemies flee:

*'Then I sent the hornet before you and it drove out the two kings of the Amorites from before you, but not by your sword or your bow."*
(Joshua 24:12)

**Pest Control**

## What they like to eat.

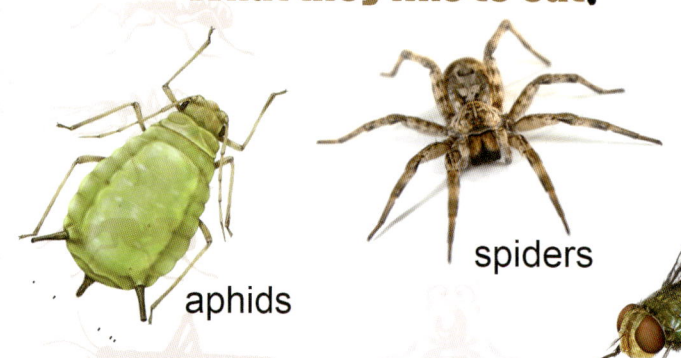

aphids

spiders

flies

**Pollinating Plants**

## Wasps and Figs

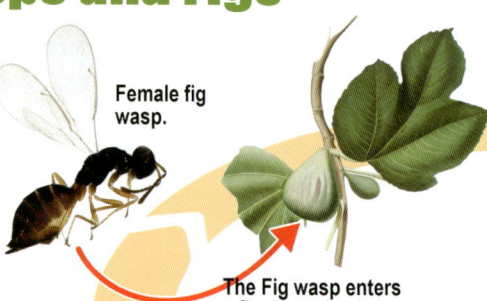

Female fig wasp.

Figs contain both male and female flowers inside.

The Fig wasp enters a fig through a hole in the bottom.

The female wasp lays its eggs in the flowers inside the fig. While doing this it pollinates some of the female flowers.

The females emerge and look for a fig to pollinate.

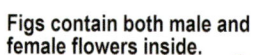

The eggs produce larve, baby male and female wasps. They grow in special pods called galls.

The male wasps then bore holes out of the figs walls so the females can leave. Then the males die.

Male wasps are born first and they fertilize the female galls.

There are 100,000 species of wasps that use other insects as hosts to lay their eggs.

Hundreds of species make paper houses.

Yellow Jacket compared to the Asia Murder Hornet.

caterpillars

beetles

crickets

A world without wasps would be a world where our crops and gardens would be overrun with pests. Some species of wasps are used to control agricultural pests.

Most of the 1000 species of tropical fig trees has its own specific fig wasp that pollinates it.

Every detail in Creation is designed to benefit other parts.

As helpful as wasps are in the garden, you are best to leave them alone.

**Wasp stinger, ouch!**

# THE SHOEBILL
## world's scariest bird

**ENDANGERED SPECIES**

*Don't tell me to have a nice day!*

5 foot tall

The Shoebill was named after a Dutch clog shoe .

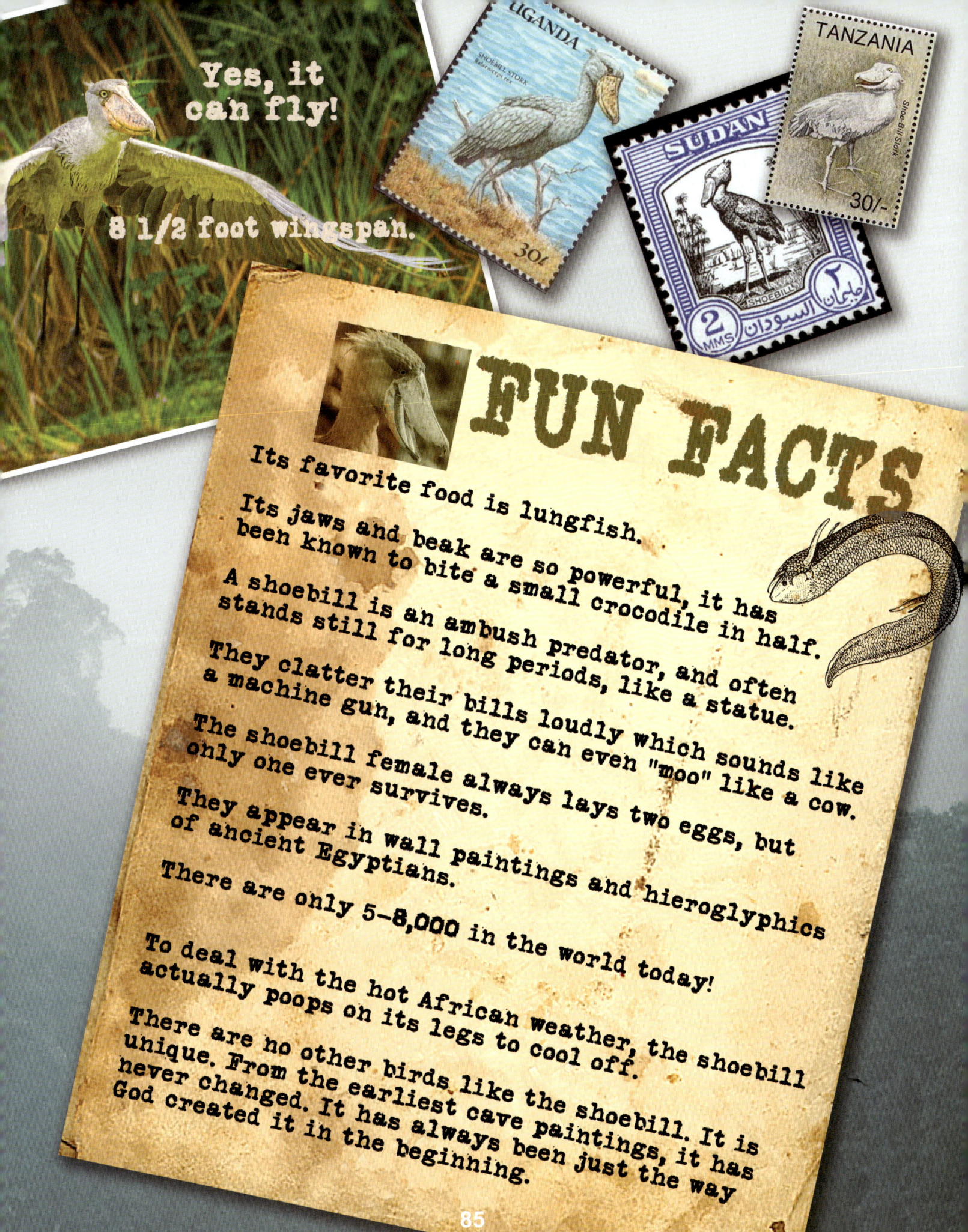

Yes, it can fly!

8 1/2 foot wingspan.

UGANDA

SUDAN

TANZANIA

# FUN FACTS

Its favorite food is lungfish.

Its jaws and beak are so powerful, it has been known to bite a small crocodile in half.

A shoebill is an ambush predator, and often stands still for long periods, like a statue.

They clatter their bills loudly which sounds like a machine gun, and they can even "moo" like a cow.

The shoebill female always lays two eggs, but only one ever survives.

They appear in wall paintings and hieroglyphics of ancient Egyptians.

There are only 5–8,000 in the world today!

To deal with the hot African weather, the shoebill actually poops on its legs to cool off.

There are no other birds like the shoebill. It is unique. From the earliest cave paintings, it has never changed. It has always been just the way God created it in the beginning.

# THE MANTIS SHRIMP
## most feared of predators

Mantis shrimp are no stranger to world records. They are famous for the fastest punch on the planet. The arm can accelerate through water up to 10,000 times the force of gravity, creating a pressure wave that boils the water in front of it. It impacts its prey with the force of a .22 calibre rifle bullet.

Their front legs use a system of biological springs, latches, and levers to power their powerful punches, enabling them to strike much more swiftly than would be possible with muscle power alone.

They are not good pets. Their kick can shatter aquarium glass.

86

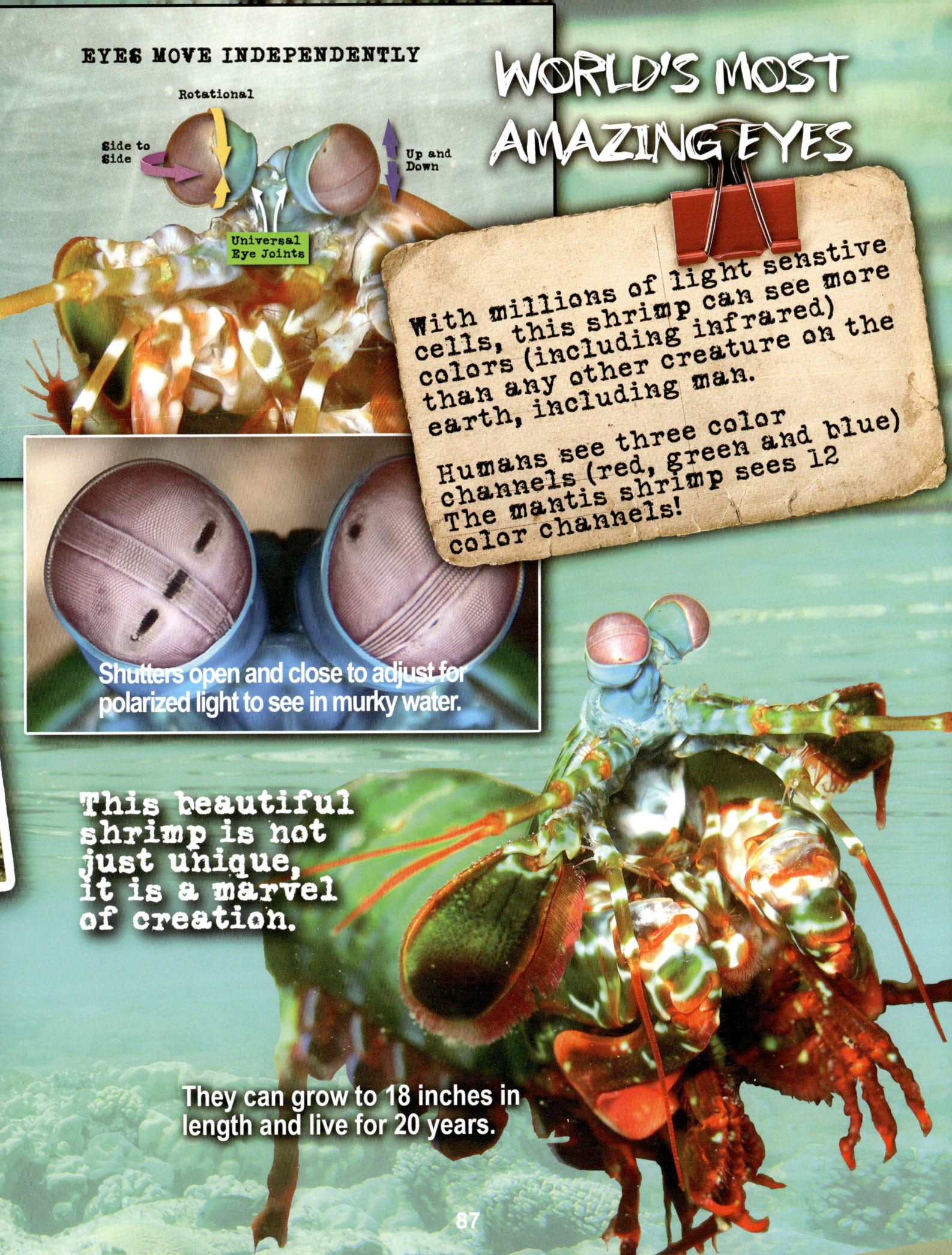

**EYES MOVE INDEPENDENTLY**

Rotational

Side to Side

Up and Down

Universal Eye Joints

# WORLD'S MOST AMAZING EYES

With millions of light sensitive cells, this shrimp can see more colors (including infrared) than any other creature on the earth, including man.

Humans see three color channels (red, green and blue) The mantis shrimp sees 12 color channels!

Shutters open and close to adjust for polarized light to see in murky water.

This beautiful shrimp is not just unique, it is a marvel of creation.

They can grow to 18 inches in length and live for 20 years.

# SWALLOWTAIL BUTTERFLY
## original "Transformer"

Only moths spin cocoons. A butterfly has a chrysalis which exists inside a butterfly caterpillar. It emerges during the process of metamorphosis. The caterpillar sheds its outer skin and reveals the chrysalis which can look like a leaf or a twig hanging by a slender thread.

**1**

### Adult Butterfly Stage
4 to 10 days

The Chrysalis hardens into a tough outer shell.

10 to 20 days

### Chrysalis (pupal) Stage

**4**

# Over 550 species

## 2

### Egg Stage
3 to 4 weeks

The eggs resemble bird poop and emit a foul odor.

Who wants to eat bird poop?

## Caterpillar (larval) Stage
10 to 20 days

One of the most unique defense systems in the world.

## 3

A bird lays an egg, it hatches and a chick is born and it flies away. That in itself is a miracle. If you agree that is a miracle, what would you call the Swallowtail butterfly?

The Swallowtail butterfly lays about 400 eggs. They resemble bird poop. No preditor bothers them.

The eggs hatch into a caterpillar that looks like a scary green snake with huge eyes. No preditor bothers them.

The caterpillar sheds its outer skin revealing an amazing thing. It has a different shell under its skin that looks exactly like a twig and it hangs by a thread from the tree. No preditor bothers them.

That shell, called a Chrysalis, becomes hard and protects what happens next. The chrysalis dissolves on the inside and transforms into a beautiful creature that emerges from the Chrysalis and flies off. It is the Swallowtail Butterfly. It now has one week to lay its eggs before it dies. The cycle begins again.

# ZOMBIE ANTS

## AND THE MUSHROOM THAT CONTROLS THEM

This is the typical growth cycle of a mushroom.

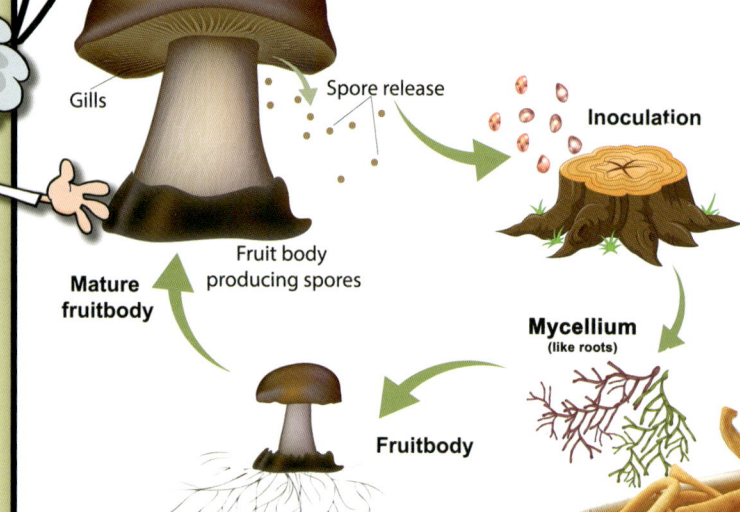

**MUSHROOM LIFE CYCLE**

Gills

Spore release

Inoculation

Fruit body producing spores

**Mature fruitbody**

**Mycellium** (like roots)

Fruitbody

However, this story is not typical!

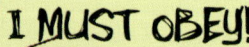

**Carpenter ant**

Ophiocordyceps mushrooms
(No, these are not french fries.)

Our story starts here.

**1** A carpenter ant in the forest eats spores from the mushroom.

**2** The fungus penetrates the ant's outer shell. The fungus grows inside the ant and hijacks its body and brain.

I MUST OBEY!

A single spore from one kind of mushroom fungus is consumed by one kind of ant. The spore contains a chemical that penetrates the ant's body armor, and then travels to the brain of the ant. The brain is reprogramed to make the ant climb a tree and sacrifice itself, so the mushroom can continue to exist. Does that sound like something that can happen by accident?

A mushroom grows out of the dead zombie ant's head, and drops its spores to the ground where more ants will eat them and become zombies.

**6.**

The zombie ant bites down on the branch clamping itself to the branch, where it dies.

**5**

SPORES

It finds a branch 10 to 12 inches above the ground.

**3** The ant's newly programmed brain tells it to find a nearby plant to climb.

**4**

12"

# THE HUMAN EAR
## the miracle of hearing

The ear is so complex and contains so many critical components that it could not have happened by accident. If one part is broken or missing, we would all be deaf. It is all or none. How could anything so balanced and perfect come about by random mutation? It couldn't. God designed and created our incredible ears.

"the hearing ear and the seeing eye, the Lord has made them both."

Proverbs 20:12

Have you ever wondered how your ears work? Sound travels on sound waves. These waves enter the ear canal causing the eardrum to vibrate. There are three small bones behind the eardrum, which by the way, are the smallest bone in your body!

These tiny bones act like an amplifier of the sound waves and transfer the vibrations to the cochlea. The cochlea is the thing that looks like a snail. Inside the cochlea is a chamber coated with tiny hairs. The entire space is filled with a special fluid. The tiny hairs are finely tuned cells that pick up different sound vibrations (frequencies).

The cochlea is a very complicated mechanism. The vibrating fluid inside interacts with the special hairs and can transform the vibrations into electrical signals. The signals are picked up by special nerve cells called auditory nerves which transfer the different frequencies to the brain. The brain sorts them all out so we understand music, human speech, or whatever sound waves enter the outer ear. How our brains process the signals and interpret them is another amazing story. It is miraculous.

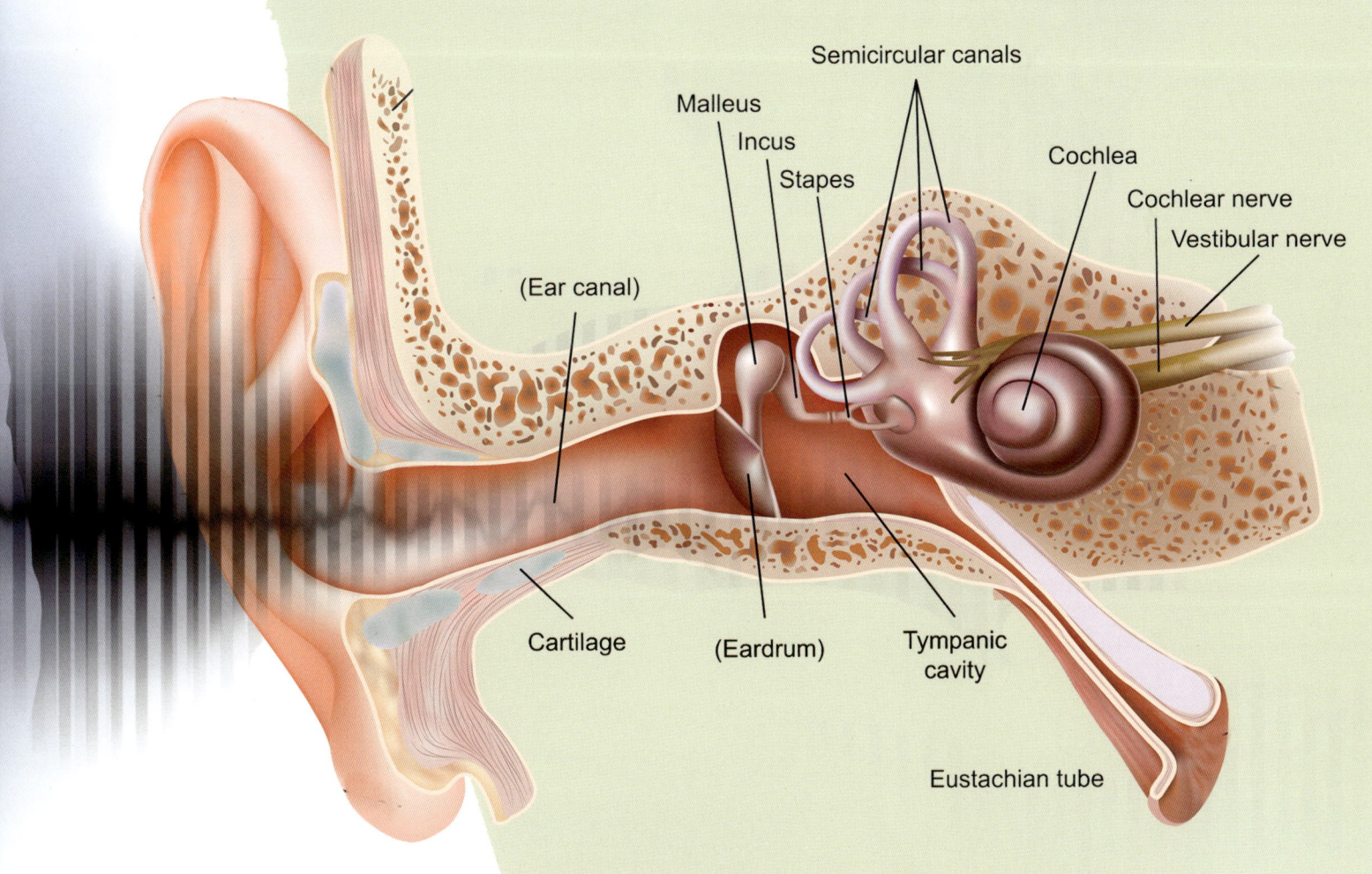

Semicircular canals

Malleus

Incus

Stapes

Cochlea

Cochlear nerve

Vestibular nerve

(Ear canal)

Cartilage

(Eardrum)

Tympanic cavity

Eustachian tube

# SPINY LONG-HORNED SPIDER

## webs are stronger than steel

Of the 45,000 species of spiders, this one is very unique. The two large horns growing out of its back can have spans up to 10 inches, more than twice the length of their bodies.

They are orb-weavers. They make the circular webs which require a large degree of accuracy and engineering.

**Found mostly in Asia**

## BREAKING NEWS

Scientists are currently combining spider web silk with goats milk to weave a nearly indestructible fiber. This amazing material can be used to repair or replace human ligaments and bones.

"In the beginning God created . . . "all the creatures that creep along the ground according to their kinds. God saw that it was good."

Genesis 1:1, 25

Only the female has the long horns!

# While all spiders spin silk, they use the silk in different ways.

- Many spin different types of webs to trap their meals.
- Some hunt for their food and use silk for making wind-sailing balloons, egg sacs or tiny "houses" to hide in.
- Some build fascinating traps and tools.
- Some produce nets to throw and some catch oxygen bubbles to carry underwater for breathing.
- Others make web slingshots and special leaf pockets for catching frogs.
- Some have devised silk pulleys to lift heavy creatures.
- They make nests and cocoons.
- Some use their silk-like hot air balloons to float away on the wind to escape predators.

The way spiders make silk is amazing. They create a liquid protein in a special gland, and use their legs to pull it out of their bodies through a set of tiny tubes called spigots.

## Silk production factory in the hind end of the spider

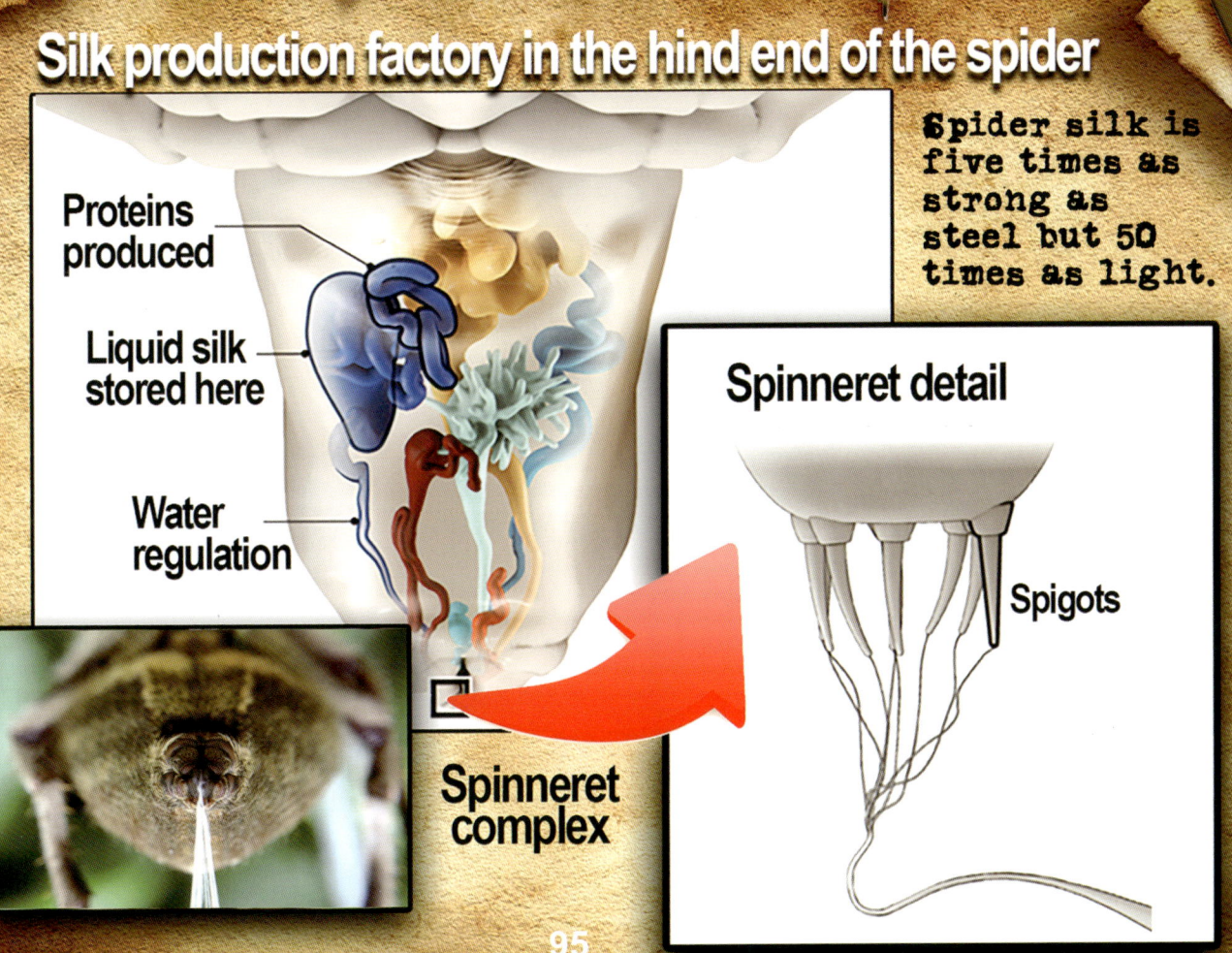

Proteins produced

Liquid silk stored here

Water regulation

Spinneret complex

Spider silk is five times as strong as steel but 50 times as light.

Spinneret detail

Spigots

# PANTHER CHAMELEON
## a coat of many colors!

Endangered Species 39c
United Nations
Furcifer pardalis (Panther Chameleon) John D. Dawson (2006)

### How and why do they change color?

So why would a chameleon want to change colors? Scientists believe that chameleons change color to reflect their moods. By doing so, they send social signals to other chameleons. For example, darker colors tend to mean a chameleon is angry. Lighter colors might be used to attract mates. Sometimes a bright color can mean it is happy lying in the sun.

### Scientific explanation.

Recent studies have shown that chameleons have a special layer of cells — called iridophores — under their skin. These special cells, which contain pigment and reflect light, are made up of hundreds of thousands of guanine crystals. Chameleons can relax or excite their skin, causing these special cells to move and change structure. Researchers found that, when this happens, these cells act like prisms, reflecting different wavelengths of light to create the variety of tones we see.

### Simple language explanation.

Just as a prism changes white light into all the colors of the rainbow by bending light, so the amazing chameleon has thousands of special cells on its skin that contain crystal-like parts that reflect light. The chameleon can actually control each cell individually to manipulate the crystal materials to change how each cell reflects light.

They are found mostly on Madagascar, an island near Africa.

They are not good pets since they like being left alone and eat 50 insects a day!

Each eye moves independantly. They can rotate 180 degrees right to left and top to bottom. Their 360 degree vision means they can see anything around them without moving their head.

There are over 200 species of chameleons in the world.

A unique blend of design and engineering, It is the work of a Great Artist. This lizard did not happen by accident.

Each skin cell acts like a prism.

## Interesting facts

In addition to the ability to change color, chameleons have many other characteristics that make them special. This includes parrot-like feet, eyes that can look in two different directions at once and long tongues and tails. On average, a chameleon's tongue is roughly twice the length of its body. In humans, that would be a tongue about 10 to 12 feet.

They can run up to 21 miles per hour. A panther chameleon is largely considered to be the chameleon that changes color the most. They can change their color in 20 seconds.

# THE PISTOL SHRIMP
## and its seeing-eye ~~dog~~ fish.

**Pistol Shrimp**　　**Goby Fish**

**ONE LARGE CLAW**

The nearly blind shrimp digs the burrow and the Goby stands guard. They both live there together for life.

The shrimp's antenna

The shrimp uses the Goby like a blind person uses a Seeing Eye dog. Whenever the shrimp is outside its burrow, it keeps one antenna on the Goby.

## The Odd Couple

The shrimp stays hidden inside the burrow if the Goby should temporarily swim away. When danger approaches, the Goby signals and disappears inside the burrow. The shrimp is right behind him.

They need each other. It is a well-balanced design.

## The claw is used like a weapon.

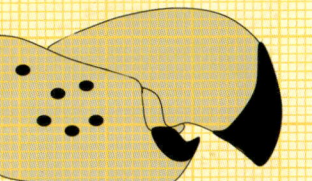

**Point**

water fills the chamber

**Set the trigger**

**Fire**

Upper claw snaps shut compessing the water, shooting a steam bullet.

As hot as the surface of the sun

**AMAZING FACT:** If the large claw is ever lost in a battle, the shrimp grows the smaller claw into a large weapon claw like the one that was lost! It is a miracle of design in creation.

## WHY IS IT CALLED THE PISTOL SHRIMP?

The shrimp's large claw is a powerful weapon. It opens and then with lightning speed clamps down on water that is trapped in a special-shaped chamber. The water compresses so much it explodes into a steam bubble with such pressure it sounds like a rifle shot.  That is where the name comes from. The steam explosion kills its prey with temperatures reaching 8,500 degrees Fahrenheit!

Some species of pistol shrimp rival whales for the noisiest creature in the ocean — pretty impressive given the size difference! The bubbles  the pistol shrimp produce can reach up to 218 decibels. Just how loud is that? Louder than a jet airplane taking off!

# THE WOODPECKER
## never has headaches

They peck at the rate of 22 times per second. It is as violent as a jackhammer. The force is ten times what would kill a person.

They stab their beaks into a tree 12,000 times a day.

Besides their eyelids, woodpeckers have a membrane that closes each time their beak strikes the wood.

A thick, spongy bone is a shock absorber between the base of the beak and the skull.

## UNIQUE BONE STRUCTURE

Very hard, outer bone layer on the beak.

Flexible Hyoid bone with fluid surrounding it for added shock absorption.

Spongy, skull bone to absorb shocks and vibrations from the pecking.

100

**The woodpecker has features found on no other bird.**

The Hyoid bone - Woodpeckers have a special bone that acts like a seat-belt for its skull. It's called the hyoid bone. It wraps all the way around a woodpecker's skull protecting its delicate brain when it pecks.

The tongue goes through a channel in the Hyoid bone.

The claws of the woodpecker are designed to hold the bird securely on a tree during its violent pecking.

The Hyoid bone is a strong, flexible bone covered in muscle that allows the woodpecker to extend its tongue out of its beak to grab food.

THE TONGUE OF A WOODPECKER CAN EXTEND MORE THAN THREE TIMES THE LENGTH OF ITS BILL!

# THE SEAHORSE
## the first submarine

### How a submarine goes underwater and returns to the surface.

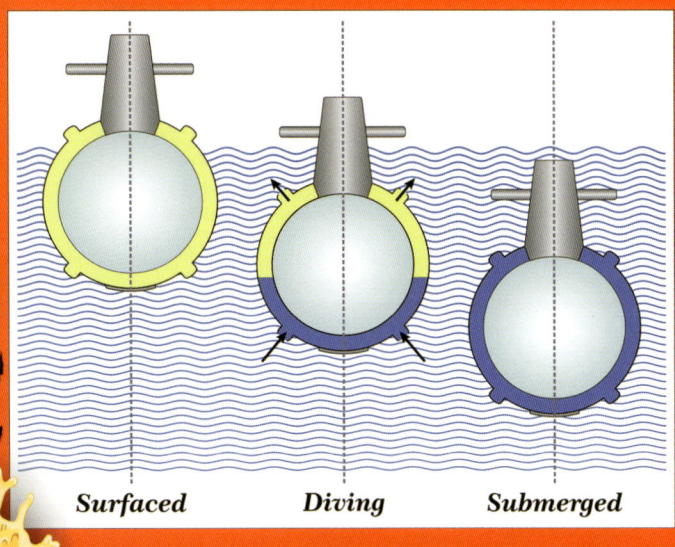

**Surfaced**  **Diving**  **Submerged**

The submarine has a hollow space in its shell. When filled with air, the submarine floats. If water is allowed to fill the space, it sinks.

The seahorse has a swim bladder in its body. It can allow the bladder to fill with water to go down or it expels the water by filling the bladder with gas making the seahorse rise, just like the submarine.

☐ There are about 40 known species of the seahorse.

☐ Seahorses prefer to swim in pairs with their tails linked together.

☐ They are the only fish that swim upright, and they avoid predators by mimicking the color of underwater plants.

☐ Few marine predators eat the seahorse, it is too bony and undigestible.

☐ Seahorses have no teeth and no stomach. Food passes through their digestive system so quickly they must eat almost constantly to stay alive. They consume 3,000 or more brine shrimp per day. Their long snout acts like a vacuum cleaner sucking in their meals.,

☐ The male carries the unborn young. The female deposits her eggs into the male's brood pouch where they remain until they hatch. In all of the natural realm, it is very rare for the male to carry the young and raise them.

☐ They can release as many as 1,500 young seahorses at once when they are grown enough to swim on their own.

They steer using two pectoral fins located near the back of the head.

Swim bladder

Seahorses propel themselves by using a dorsal fin on their back.

Brood pouch where fertilized eggs mature and then are released as baby seahorses.

"Speak to the earth, and have it teach you; And have the fish of the sea tell you. Who among all these does not know that the hand of the Lord has done this."

Job 12:8, 9

The leafy seadragon is part of the seahorse family.

Seahorses are poor swimmers so they anchor themselves with their tails to sea grasses during heavy currents.

# THE CLOWNFISH

## needs the sea anemone to survive

## The Sea Anemone

The sea anemone looks like a plant but is a marine animal. Its tentacles are poisonous. The sting contains a neurotoxin that paralyzes its prey. They mostly feed on plankton, crabs, and fish, and sometimes jellyfish.

## The Clownfish

The clownfish has a protective mechanism for the sting of the anemone. When a clownfish has been stung it produces a protective coating on its skin which allows it to swim around within the tentacles without harm. The small clownfish uses the sea anemone as a place of safety from predators.

There are 30 different species of clownfish.

PILIPINAS P7

2010

*Dilubre*
©
*Photos*

True Clownfish
(*Amphiprion percula*)

## Mutually Beneficial

The clownfish helps the sea anemone by consuming the parasites which threaten it. The clownfish provides nutrients through its feces (poop). The movement of the clownfish through the tentacles helps oxygenate the anemone. Both the anemone and the clownfish live in harmony, they are mutually beneficial to each other.

This is one of many pieces of evidence of a Grand Designer who created them both to survive together.

**Distribution of clownfish and sea anemone**

Asia

Africa

French Polynesia

Australia

There are more than 1,000 species of sea anemones.

# KOMODO DRAGON
## world's largest lizard

**DID YOU KNOW?**

*lindungilah margasatwa*
**75** *sen*
*komodovaran*
**Republik Indonesia**

- ☐ Komodo dragons are found on the islands of Indonesia.
- ☐ A Komodo dragon can grow to 10 feet, and can weigh over 300 pounds.
- ☐ A Komodo dragon uses its tongue to smell a dead animal 3 to 6 miles away.
- ☐ A Komodo dragon can swim like a crocodile, dive up to 15 feet deep, and run as fast as 12 mph.
- ☐ A Komodo dragon has 60 sharp, curved, and serrated teeth. It has two venom glands in the lower jaw.
- ☐ A Komodo dragon can swallow a goat-sized animal whole because it has a loosely articulated jaw, flexible skull, and an expandable stomach.
- ☐ A Komodo dragon can eat as much as 80% of its weight in one feeding.
- ☐ The Komodo dragon could survive on 12 big meals per year.
- ☐ A Komodo dragon has a small tube underneath its tongue that connects directly to its lungs. This allows the dragon to breathe while swallowing its food.

**Gulp!**

Komodo dragons kill and eat young komodo dragons, which is about 10% of their diet.

In the beginning God created a beautiful world. It was perfect. Komodo dragons didn't eat cute goats. People didn't hurt people. But that all changed when man, God's crowning creation, decided to rebel against His Creator. God cursed all of His creation. It has been broken since. The good news is that God sent His Son to redeem us and everything else from the curse.

"We know that the entire creation is groaning together, and going through labor pains together, up until the present time." (Romans 8:22)

One day in the future this dark chapter will come to an end and God will remake it all. The new heavens and new earth will be restored to perfection and there will never be a curse again. The lion will lay down with the lamb. You may even have a pet komodo dragon on that day. Jesus will make all things new. All who have trusted in Jesus will experience an eternal new day.

ENDANGERED SPECIES

Unless you are a top olympic sprinter, you cannot outrun a komodo dragon!

# THE HONEY BEE

## Part 1 — in the field

### THE TEAM

| Queen bee | Drone bee | Worker bee |
|---|---|---|

1. The queen bee is the only female in the hive that lays eggs and will lay about 600 – 1,500 daily. The queen bee can designate if a newborn bee is to be a male or female bee.

**LIFESPAN 2-7 years**

2. Drone bees are male bees without stingers whose primary purpose is to mate with the queen bee.

**LIFESPAN 3 months**

3. Worker bees are female bees who collect the nectar and pollen, build the honeycomb, make the honey, feed the larvae bees, repair the hive, and care for the queen. In other words, the female bee does all the work.

**LIFESPAN 40 days**

## THE BUZZzzz

God designed all of His living creations to reproduce, including flowers. Some flowers need help from a little insect called the honey bee. Many flowers produce a sugary liquid called nectar to entice the honey bee and other insects. The honey bee needs the nectar and pollen as a necessary food source.

The bee visits the flower to suck up the nectar with its straw-like mouth, collects pollen from the male part of the flower, and carries it in pollen baskets on the bee's back legs. The bee then goes to another flower to get more nectar, and some of the pollen is deposited on the female part of the other flower, thus pollinating it.

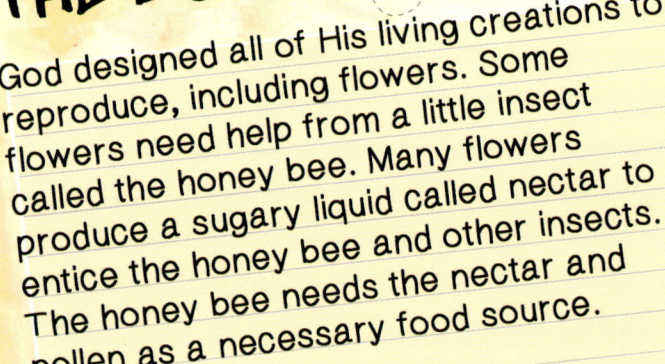

### BEES HAVE FIVE EYES!

Three small eyes on top of the head are used for navigation and orientation.

Each of the two large compound eyes have 7,000 lenses! These help the bee see shapes and colors.

Antennae allow the bee to feel and smell.

Tongue reaches the nectar in the flowers.

Special glands produce royal jelly which feed a new queen bee.

Without bees, these would not exist.

**Pollination**

Pollen
Pollinator
Pollen
Pollen grains
Pollen
Stigma
Ovule
Ovary

Special honey stomach used to carry nectar back to the hive.

Glands that produce the wax to build the honeycomb.

Baskets on each leg to carry pollen back to the hive.

Honey bees pollinate 80 percent of all flowering plants, including 130 types of fruits and vegetables.

There are no evolutionary explanations for:
☐ Its honey stomach
☐ Its ability to make honey
☐ Its beeswax glands
☐ Its royal jelly glands

However, they all have to be fully functional for the bee to exist.

Turn the page for Part 2

# THE HONEY BEE

## Part 2     at the hive

Wow, I think I have too much pollen!

**Bees return from the flowers with their pollen sacks full.**

The bees return to the hive and pump the nectar into another bee's mouth adding enzymes which thickens it.

The bees fan the nectar with their wings to evaporate some of the water until the nectar is less than 18%. water. Any additional water would cause the honey to ferment and spoil.

Back at the hive, bees transfer their pollen and nectar.

**The protein-rich pollen is stored in the cells for nourishing young bees.**

**Honey is stored in honeycomb cells.**

**Cells are capped off with beeswax for future use.**

# The hexagon is nature's perfect shape.

Beeswax is produced in a gland in the abdomen of honey bees and is excreted to build the hive itself. The worker bees have eight glands in their abdomens to make the wax. The wax from the bee hardens as soon as it hits the air and forms a wax scale. Other worker bees scrape the wax, chew it with some honey and pollen, and use it to build the honeycomb.

The bee begins by building a circular cell but uses its body heat to shape it into a hexagon (six sides). The hexagon is the most efficient and strongest shape to connect other hexagons.

**Making the honeycomb.**

Honey bees repair their hive using propolis (bee glue), a mixture of pollen, beeswax, and resin collected from tree sap. Bees diligently seal cracks with propolis to prevent unwanted insects from entering the hive.

Honey is a source of food stored in the honeycomb for the winter months when there are no flowers. Bees make more than they need so that beekeepers can take some for their use. Honey bees must fly about 55,000 miles to produce a pound of honey, visiting around 2 million flowers. Bees can fly about 15 miles per hour.

God created this amazing creature to supply us with honey, fruits and vegetables. They are His gracious provision for all mankind and are a living testimony of His love for us.

Bees heat and cool their hive by fanning their wings to keep it between 93 and 95 degrees year-round.

Bees communicate with other bees where to find food by doing a "waggle dance," indicating the location and distance to find a food source.

111

ON

# THE ORIGIN OF SPECIES

## BY MEANS OF NATURAL SELECTION,

OR THE

PRESERVATION OF FAVOURED RACES IN THE STRUGGLE
FOR LIFE.

By CHARLES DARWIN, M.A.,

FELLOW OF THE ROYAL, GEOLOGICAL, LINNÆAN, ETC., SOCIETIES;
AUTHOR OF 'JOURNAL OF RESEARCHES DURING H. M. S. BEAGLE'S VOYAGE
ROUND THE WORLD.'

LONDON:
JOHN MURRAY, ALBEMARLE STREET.
1859.

*The right of Translation is reserved.*

We end this book where we began. How did it all happen? Where did we come from? It either all happened by itself or it was created.

We will summarize both Evolution and Creation.

## The Book

Charles Darwin published THE ORIGIN OF THE SPECIES in 1859. It shocked the religious community by challenging the existence of a Creator to explain the diversity of life on earth. To this day no single book has done more to damage faith in God. Darwin's work has become the standard for higher education and biology to explain the complexity of life by natural changes without the need of God.

## The Message

The theory of Evolution is described as a natural process where simple life forms slowly change into higher life forms. The change occurs by mutations in living cells. Cells that produce damaging mutations are rejected by the cells through a process called natural selection. Mutations that improve the cells remain. Over billions of years of small changes, the world and its variety of lifeforms became more and more diversified.

CHARLES DARWIN 1809 - 1882

THE ORIGIN OF SPECIES

BY MEANS OF NATURAL SELECTION,

भारत INDIA

# EVOLUTION

## The Choice

Charles Darwin made a decision and dedicated most of his life to discover how the variety of complex life forms came into existence. He traveled to remote places to find and document his findings. He also made a choice to leave God out of the story. He replaced God with natural selection, a process within cells to determine their own destiny. God, in Darwin's world, had no place. The central belief of evolution is "no belief" at all.

## The Ending

Since evolution is described as a series of accidents, there is no direction or hope in the system. We are described as advanced animals, nothing more. The entire universe began with an explosion and one day, scientists say, it will collapse back into itself and end like it began.

They believe we live and then we die and no longer exist after death. Evolution ends in a grave. It is a hopeless life.

A British cartoonist drew this caricature of Charles Darwin 10 years after his book was published. Many rejected his theory.

**Turn the page for Creation**

## EXERCISE

Take a grain of sand and hold it at arms length to cover the darkest part of a starry sky.

The area of the sky you are covering is what is seen in the photograph on the right side. Even though it is a very small part of the universe, it contains more stars than you might think.

The photo was taken by a large telescope in orbit in space called the James Webb space telescope. The photo reveals 10,000 galaxies, not stars, but GALAXIES! Each galaxy contains 100 billion stars!!

Up to now we have been looking at the earth's animals, insects and microscopic world, which by itself is incredible. Now we turn our eyes to the heavens and discover another immense and beautiful sight.

Everything in the universe is in balance and as astronomers say, it is perfectly tuned, no part is out of place. If gravity was slightly stronger or weaker, it would self destruct. Books are written on the many ways the universe is perfectly tuned. It is planned and designed, not an accident. It did not happen by itself.

The Bible tells us that the heavens declare the glory of God. With evolution there is no hope, but with God we find eternal hope.

# CREATION

"In the beginning God created the heavens and the earth."
Genesis 1:1

Each dot is a galaxy that
contains 100 bilion stars!

Photo taken by the James Webb Space Telescope

# GOD's
## wonderful plan for us . . .

## ETERNAL LIFE

**Everyone has sinned.**

**Romans 3:23**
*For all have sinned and fall short of the glory of God.*

**We deserve punishment but God offers forgiveness.**

**Romans 6:23**
*For the wages of sin is death, but the free gift of God is eternal life in Christ Jesus our Lord.*

**Christ died for our sins.**

**Romans 5:8**
*But God demonstrates His own love toward us, in that while we were yet sinners, Christ died for us.*

**We need to trust Christ as our Saviour.**

**Romans 10:9, 10**
*That if you confess with your mouth Jesus as Lord, and believe in your heart that God raised Him from the dead, you will be saved; 10 for with the heart a person believes, resulting in righteousness, and with the mouth he confesses, resulting in salvation.*

If you believe these things, pray to God and confess your sins and ask Jesus to be your Savior and promise to follow Him.

**God promises eternal life to all who call on Him.**

**Romans 10:13**
*For Whoever will call on the name of the Lord will be saved.*

With God, there is a happy ending. With evolution, there is only an ending.

We were made in God's image and He wants us to be in His forever family in a place He has just for us.

Now that is hope!